How Evolution Works
(and Why Socialism Doesn't)

Plus:
A New Way to See Life
(and *your* life!)

Richard Showstack

How Evolution Works (and Why Socialism Doesn't)

©2025

Nature is red in tooth and claw.

Tennyson

**Every death, even the cruelest death,
drowns in the total indifference of Nature.
Nature herself would watch unmoved
if we destroyed the entire human race.**

From *Marat/Sade,*
a 1967 British film adaptation
of Peter Weiss' play, *Marat/Sade*

**A chicken is an egg's way
of making another egg.**

Samuel Butler

All is vanity.

Solomon

Note

While reading David Sloan Wilson's marvelous book, *Evolution for Everyone: How Darwin's Theory Can Change the Way We Think About Our Lives*[1], I noticed that his goal in writing his book was very similar to my goal in writing this book. So, with due respect to Prof. Wilson, I am going to quote a few lines from the introduction ("The Future Can Differ from the Past") to his book:

> I use the principles of evolution to understand the world around me…. I include all things human along with the rest of life…. I and my fellow evolutionists study the length and breadth of creation, from the origin of life to religion….
>
> [M]ost people who do accept evolutionary theory don't use it to understand the world around them. For them it's about dinosaurs, fossils, and humans evolving from apes, not the current environment or human condition. The polls don't measure the fraction of people who relate evolution to their daily lives, but it would be minuscule….
>
> With respect to evolution, most scientists and intellectuals would say that they accept Darwin's theory, but many would deny its relevance to human affairs or would blandly acknowledge its relevance without using it themselves in their professional or daily lives.

Like Prof. Wilson, in this book I will try to explain evolution's relevance to our daily lives, objectively based on the facts, without emotion or bias.

---

[1] Wilson, D. S.

Table of Contents

Warning: This is not a "feel good" book!
*In the following pages, you will be exposed to some ideas that
will be new, some that may shock you, and some that may
even shake the very foundations of your beliefs! If you do not
think you can handle that, then stop here. But if you are brave
and open-minded enough to learn about, consider and
perhaps even accept some new ideas, then proceed.*

Notes:

1) Even if you believe in "Creationism" and "Intelligent Design"
   and believe that the Theory of Evolution is ungodly, please
   keep reading. I think that you will find that "nature" is pretty
   wise, too. In fact, if you mentally replace every mention of
   "Evolution" or "Nature" in the book with "God," what this book
   explains will still make sense.

2) Of course, genes do not have the capacity to make plans; they
   are just molecules. However, it helps to understand how
   evolution works if it is discussed in a "teleological[2]" way.
   Therefore, I will use words such as "purpose" in a figurative
   sense.

3) Much of the explanations in this book of how Evolution works
   are based on the book, *Nature and Man's Fate* (1961), by
   Garrett Hardin. I strongly suggest that you read it (if you can
   get your hands on a copy of it).

---

[2] relating to or involving the explanation of phenomena in terms of the
purpose they serve rather than of the cause by which they arise.

Foreword:
The "Conventional Wisdom" is *Wrong!*

People on the left side of the political spectrum in the U.S. say they "believe in evolution," but do they really understand how evolution works? And religious people on the right believe that one cannot believe in evolution and be a godly person at the same time.

And almost everyone in modern society more or less agree on certain things, for example, that it is immoral to let people die because they do not have enough resources to survive, and that everyone deserves free medical care. However, in Part Two of this book I will explain why, in terms of what we know about evolution, the "conventional wisdom" is wrong!

First I will give you a basic explanation of *how evolution works*.

It is not my intention to promote or criticize a particular political agenda, but in the second part I will discuss some *implications and ramifications* of what I explained in Part One.

Finally, I will discuss a deeper philosophical question: "If life is just about a bunch of molecules randomly building organisms, how can there be a meaning to life?"

What you read may be unpleasant, it may even make you angry or disgusted, but sometimes the truth is like that. In the following pages I am going to explain real "the facts of life" to you.

Preface

Thomas Sowell wrote:

> What are the underlying assumptions behind the
> very different ideological visions of the world
> being contested in modern times? The purpose
> here will not be to determine which of these
> visions is more valid but rather to reveal the
> inherent logic behind each of these sets of views
> and the ramifications of their assumptions which
> lead not only to different conclusions on
> particular issues but also to wholly different
> meanings to such fundamental words as
> "justice," "equality," and "power...."[3]

And Richard Joyce wrote:

> If uncomfortable truths are out there, we should
> seek them [out] and face them like intellectual
> adults, rather than eschewing open-minded
> inquiry or fabricating philosophical theories
> whose only virtue is the promise of providing
> soothing news that our heartfelt beliefs are true.[4]

Last, as Miller and Kanazawa wrote in the introduction
to their book, *Why Beautiful People Have More Daughters*
(2007):

---

[3] Sowell. T.
[4] Joyce, R.

[I]t would hardly be appropriate to criticize scientific theories simply because their implications are immoral, ugly, contrary to our ideals, or offensive to some....[T]he implications of many of the ideas we present in this book (whether ours or someone else's) are indeed immoral, ugly, contrary to our ideals, or offensive to either men or women (or some other groups of people). However, we must state them as they are because, to the best of our scientific judgment, they are true. That does not mean that we endorse all possible consequences and implications of our observations or believe that they are somehow good, right, desirable, or justifiable.[5]

Plato's Allegory of the Cave

In the nineteenth century, Ralph Waldo Emerson wrote: "Things are in the saddle and ride mankind." However, I would update that by saying: "*Genes* are in the saddle and ride mankind."

I think that Plato's "allegory of the cave"[6] may help explain how I see genetics and "objective reality."

In Plato's allegory, we humans are like people who have spent our entire lives in an underground cave chained in place so that we can see only the shadows of reality projected on a wall in front of us by a large fire behind us. Some people will be content to watch the changing patterns

---

[5] Miller, A.S. & Kanazawa, S.
[6] "The Allegory of the Cave"

of shadows in front of them while others will try to observe them and learn from them, but, in either case, they are seeing only the "projections" of reality.

However, imagine that one such human manages to break free and escape the cave. He will be able to see the true nature of existence!

He may then return to the cave and try to explain to the others that what they take for reality is, in fact, only its projection, but they are unlikely to believe him. They are more likely to be content to continue to hold dear their longtime beliefs, no matter how false they are.[i]

For the purposes of this analogy, the "other things" that we can't directly see are our genes (and how the way they work affects us). We see their "projections" in our lives, but only as they want us to see them. Each species has (only) the sensory abilities it needs to gather the information it needs from the world to maximize its chance of survival. Each of us is the repository of the cumulative abilities that our species has empirically "learned" are helpful to our survival.

We live in the reality of what we see, feel and experience—the shadows on the wall of the cave—but at the same time, within us, in billions upon billions of cells, our genes—copies of which may have existed for hundreds of millions of years or longer—are constantly going about the jobs that will help insure that copies of themselves will continue to exist in the future.

Going back to Plato's allegory of the cave, I am like the philosopher who has been freed from the cave. I was born with the "disability" of not possessing the "believe things which aren't true" gene. I may not be able to perceive the true form of reality rather than the mere shadows seen by the prisoners, but I at least understand that the true form of reality rests in our genes and that what we perceive in the world are only the "shadows on the wall" that we create in our minds in ways that are determined by our genes.

The Importance of (False) Beliefs

However, in order to survive it's also necessary for some (all?) people to believe in some things which *aren't* true. Believing some things that aren't actually true and that we have some God-given abilities can make us happier and healthier.[7] Joyce maintains, "[I]f in certain domains *false* beliefs will bring more offspring then that is the route natural selection will take every time."[8] And Kevin Foster, an evolutionary biologist at Harvard University, believes, "The tendency to falsely link cause to effect – superstition – is occasionally beneficial,"[9] and as long as the cost of believing a superstition is less than the cost of missing a real association, superstitious beliefs will be favored.

As with most other traits, the best adapted people have a *moderate* propensity to believe things which aren't true because such a trait makes it more likely that they will survive and pass their genes on to the next generation. In terms of "survival of the fittest," it is (in most cases) not

---

[7] People on the left of the political spectrum make of fun of people who believe in "silly myths" of religions, but this just shows how little people on the left understand Evolution (and religion). I will discuss this in more depth when I discuss religion later.

[8] Joyce, R.

[9] Callaway, E.

good to be too far on one side of the scale—of being a total skeptic like I am—nor is it good to be too far on the other side of the scale—of believing that there are extraterrestrials, conspiracies and government plots all around us.

Thus, we thus live in two realities, one of which is unseen to us. One of those realities unfolds before us in "human time" — second by second, day by day, year by year, lifetime by lifetime. The other (genetic) reality unfolds in "biological time" over generations, over hundreds, thousands, tens of thousands, hundreds of thousands, millions — even billions —of years.

Plus, we *see* and *hear* and *feel* and *taste* and *smell* and *interpret* the world and *form our beliefs about it* in ways that our species' evolutionary experience has determined are the best ways for us to *see* and *hear* and *feel* and *taste* and *smell* and *interpret* and *form our beliefs about* the world *in order to carry out the function of our genes*, and those beliefs about the world may or may not have any close relation to "objective reality."

My Goal

In the following pages, my goal (as was well stated by Thomas Sowell above) is not "to reconcile visions or determine their validity, but to understand what they are about, and what role they play in political, economic, and social struggles." In other words, this book is meant to be *descriptive*, not *prescriptive*.

I also want to emphasize that in discussing the way Evolution works I am not saying that it is a good thing that it works that way. I am merely the messenger, reporting the facts.

I also don't support those who attack Evolution as impossible (or at least amoral). Actually, I find it ironic that those who claim most vehemently to believe in God do not appreciate the vastness, age and complexity of "God's" universe.

In fact, I believe that if you understand genetics, natural selection, etc., you can understand much about why individuals—and groups of people—are the way they are and why they act the way they do. If you come to truly understand how Evolution, Natural Selection, Adaptation, etc., work, it will change your way of seeing life and change your view of social reality.

Unfortunately, we are not always aware of how nature is influencing us: Like all other animals, our genes quietly influence us in ways that get us to act in ways to preserve them.

Of course, that doesn't mean that we need to be nature's slaves:

> [N]ature isn't a moral authority, and we needn't adopt any "values" that seem implicit in its workings—such as "'might makes right." Still, a true understanding of human nature will inevitable affect moral thought deeply and…legitimately.[10]

---

[10] Wright, R.

Darwin's theory of natural selection is like a recipe with three ingredients.

As Wilson explains in *Evolution for Everyone*:[11]

> We start with variation. Individuals such as you and I differ in just about anything that can be measured, such as height, eye color, or quickness to anger. Then we add consequences. The differences between you and me sometimes make a difference in our ability to survive and reproduce. Perhaps your superior size enables you to take my stuff or even kill me directly. Perhaps my inferior size enables me to survive the winter on less food. The details depend up on our particular traits and the environments we inhabit. The final ingredient, a sort of yeast that makes the recipe come to life, is heredity. For many traits, offspring tend to resemble their parents.

Four Things You Need to Understand About Evolution

Before we get into the details of my argument, there are four things you need to understand:

1) All living things (including you and me) are constructed by genes. The genes do not know what they are doing — they do not have consciousness — but if they do their job correctly, they will help to construct and maintain structures (plants or animals) that will survive and pass

---

[11] Wilson, D. S.

copies of the genes on to other similar "structures" in the future.

2) "Your genes reside in you because they had a net positive effect averaged across all of the individual organisms and environments they have inhabited over thousands of generations."[12]

3) Nature *doesn't care*! She doesn't care about pain, suffering, sadness, anger, love, etc., *except* to the extent that these emotions help us accomplish #1 above.

4) The best way to understand a species—any species— is by observing them and then making some hypotheses based on our observations, and this includes *Homo sapiens*.

> "For the moment, we need to adopt the detached perspective of the Greek Gods, watching the comedy and tragedy of life below."[13]

In other words, we should ignore the explanations that people give for their actions because such explanations can be self-serving or simply wrong.

Because of the four things listed above, in the following pages I am going to try, as objectively as I can, to describe how the world works from the point of view of *genes*, which means it's not about "right" and "wrong" and it's not about "better" or "worse." For genes, it's simply about cause and effect—what works and what doesn't work.

---

[12] Wilson, D. S.

[13] Wilson, D. S.

It is ironic that all living things are simply the creations of molecules whose "actions" are entirely governed by the physical laws of cause and effect. Genes don't think about what they do. They don't suffer or feel regret when they make a mistake (or when, as a result of their actions, a person—or millions of people— suffer). But *we should care* how genes (and evolution, adaptation and natural selection) work.

I also want to make clear that I am not a scientist and this is not a pure "science" book nor is it a book of "political science." Rather, it is a polemic. What I will do is *attempt to use scientific observations of what Evolution tells us about how life—and Evolution—work to examine how society works and how well various social policies work.*

What's to Come

In **Part One**, I will explain *how Evolution works.*

In **Part Two**, I will consider some *implications* and *ramifications* of what was been discussed in Part One.

In **the Conclusion**, I will discuss what it all means for us and then I will give some *final thoughts.*

I don't expect any changes in social policies to occur as a result of the publication of this book. All I hope is that, as a result of reading this book, you will come to a new and deeper understanding and appreciation of how the forces of Evolution and Adaptation interact with human beliefs and social policies.

I expect criticism from both sides for the issues I raise in the book, but, as William Harvey wrote:

> I tremble lest I have mankind at large for my enemies, so much doth wont and custom become a second nature. Doctrine once sown strikes deep its root, and respect for antiquity influences all men. Still the die is cast, and my trust is in my love of truth and the candour [sic] of cultivated minds.[14]

---

[14] On the Motion of the Heart and Blood in Animals: William Harvey

Part One
How Evolution Works

How Life Started and Evolved

The earth formed about 4.5 billion years ago, and recent research suggests that there were water molecules on earth from the very beginning of its existence. However, the first evidence of life appeared about 3.8 to 4.0 billion years ago, about 500 to 700 million years after the earth formed.

Five hundred million years is a long time, especially considering that chemical reactions can take place in an instant. There are 31,536,000,000,000 seconds in a million years, so there are 15,768,000,000,000,000 seconds in 500 million years. That's a lot of seconds! If you multiply that number by the number of molecules that were reacting with each other every moment on earth, you get a number so large that it is almost beyond human comprehension.

So for 500 million years molecules were bumping into each other, sometimes combining to form larger molecules. Then, one day, something amazing happened: By chance, two molecules collided and joined together to form a new molecule which itself starting making exact copies of itself!

Now, although it probably did it in a fraction of a second, let's say it did this once a second. And, since each of these new molecules was an exact copy of the "parent" molecule, they too started making new copies, which started making new copies, which started making new copies of themselves, etc., etc., etc.

As you can see, it would not take long for there to be millions, billions, trillions of these new molecules!

But, as anyone who has used a photocopy machines knows, copying is not an exact science; sometimes mistakes happen. Let's say that a mistake did happen (an atom was removed or added or replaced by another atom) in this copying process about one percent of the time. (Scientists estimate that humans are born with, on average, about 200 mutations in their DNA, which is about one percent of the approximately 20,000 genes we have.)

In the vast majority of instances (let's say, 99 percent of the time) these changes *hurt* the molecule's ability to remain in one piece and make copies of itself, so 99 percent of the altered molecules would have a decreased likelihood of making copies of themselves.

However, in one percent (that is, one in a hundred) of the replications, the change made the resulting molecule *better able* to survive and make copies of itself.

Now since this process of mutation happened only one percent (one time in a hundred) of the time and only one percent (one in ten thousand) of the changes were beneficial, every time a new "generation" of molecules was born only *one in ten thousand* of them "evolved" to be better adapted to its environment than the its "parent" molecule.

However, when billions and trillions of new molecules are being produced *every second*, that means that thousands of "better adapted" molecules are also being produced every second! So, after a very short time, there were a large number of new varieties of molecules being produced, and occasionally (let's say, one in a trillion times) two of these new molecules would "meet" and "get married" (form a new complex molecule) and start having "children" (making copies of themselves). And remember that, about one percent of the time, one of the copies would have a "mistake" (a mutation), and almost all of these mistakes would hurt the new molecules' ability to make copies of themselves, but once in a million times…and so on and so on.

But these new molecules were still not "alive," for:

> Life is a characteristic that distinguishes physical entities that have biological processes (such as metabolism, reproduction, signaling and self-sustaining processes that respond to stimuli) from those that do not… [organisms that do not have these biological processes] are classified as inanimate.[15]

Again, also according to Wikipedia.org, living organism are:

> open systems that maintain homeostasis [the tendency toward a relatively stable equilibrium between interdependent elements, especially as maintained by physiological processes. originating from within the organism] are composed of cells, have a life cycle, undergo metabolism, can grow, adapt to their environment, respond to stimuli, reproduce and evolve.[16]

Last:

> In biology, an organism is any living system (such as animal, plant, fungus, or micro-

---

[15] Hayden, T.
[16] Life

organism). In at least some form, all organisms are capable of response to stimuli, reproduction, growth and development, and maintenance of homeostasis as a stable whole.[17]

So just having the ability to copy itself does not make a molecule "alive."

These molecules were instead "protocells": "self-organized, endogenously ordered, spherical collections of lipids."[18] However, after these protocells banged into each other for another 300 to 500 million years or so (until around 3.5 billion years ago), the first single-celled organisms appeared.

Then, it took almost *three billion years* (around *100,000,000,000,000,000 seconds!*) for the first multicellular organism to appear (around 600 million years ago).

According to newscientist.com:

It is unclear exactly how or why this happens, but one possibility is that single-celled organisms go through a stage similar to that of modern choanoflagellates: single-celled creatures that sometimes form colonies consisting of many individuals.[19]

---

[17] Marshall, M.
[18] "Biology: Study of Life"
[19] Than, K.

All life on Earth evolved from a single-celled organism that lived roughly 3.5 billion years ago.

Using computer models and statistical methods, biochemist Douglas Theobald calculated the odds that all species from the three main groups, or "domains," of life evolved from a common ancestor—versus, say, descending from several different life-forms or arising in their present form, Adam and Eve style.

The domains are bacteria, bacteria-like microbes called Archaea, and eukaryotes, the group that includes plants and other multicellular species, such as humans.

The "best competing multiple ancestry hypothesis" has one species giving rise to bacteria and one giving rise to Archaea and eukaryotes, said Theobald, a biochemist at Brandeis University in Waltham, Massachusetts.

But, based on the new analysis, the odds of that are "just astronomically enormous," he said. "The number's so big, it's kind of silly to say it"—1 in 10 to the 2,680th power, or 1 followed by 2,680 zeros.

Theobald also tested the creationist idea that humans arose in their current form and have no evolutionary ancestors. The statistical analysis showed that the independent origin of humans is "an absolutely horrible hypothesis," Theobald said, adding that the probability that humans were created separately from everything else is 1 in 10 to the 6,000th power.[ii]

Then it took *another* 300 million years for some of the multicellular organism to acquire bilateral symmetry.

[About 590 million years ago] the Bilateria, those animals with bilateral symmetry, undergo a profound evolutionary split. They divide into the protostomes and deuterostomes. The deuterostomes eventually include all the vertebrates, plus an outlier group called the *Ambulacraria*. The protostomes become all the arthropods (insects, spiders, crabs, shrimp and so forth), various types of worm, and the microscopic rotifers.[20]

In other words, humans appeared only after "life" had been evolving for almost 3.5 billion years. Then it took *another 100 million years* for the tiny "Bryozoa" to appear (about 480 million years ago).

Bryozoa …are a phylum of aquatic invertebrate animals. Typically, about 0.5 millimeters (0.020 in) long, they are filter feeders that sieve food particles out of the water using a retractable lophophore, a "crown" of tentacles lined with cilia.[21]

Our unique attributes evolved over a period of roughly 6 million years. They represent modifications of great ape attributes that are roughly 10 million years old, primate attributes that are roughly 55 million years old, mammalian attributes that are roughly 245 million years old, vertebrate attributes that are roughly 600 million years old, and attributes of nucleated cells that are perhaps 1,500 million years old. If you think it is unnecessary to go that far back in the tree of life to understand our own attributes, consider the humbling fact that we share with nematodes (tiny wormlike creatures) the same gene that controls appetite.

[20] Than, K.
[21] Bryozoa

At most, our unique attributes are like an addition onto a vast multiroom mansion. It is sheer hubris to think that we can ignore all but the newest room.[22]

What is "speciation"

New species arise through a process called "speciation." Speciation is the formation of new and distinct species in the course of evolution. Speciation involves the splitting of a single evolutionary lineage into two or more genetically independent lineages which can no longer reproduce with each other.

> Darwin envisioned speciation as a branching event. In fact, he considered it so important that he depicted it in the only illustration of his famous book, *On the Origin of Species*.[23]

How do new species arise?

So how does speciation occur?

> For speciation to occur, two new populations must be formed from one original population, and they must evolve in such a way that it becomes impossible for individuals from the two new populations to interbreed.[24]

And:

---

[22] Wilson, D. S.
[23] Species & Speciation
[24] Species & Speciation

Speciation can be driven by evolution, which is a process that results in the accumulation of many small genetic changes called mutations in a population over a long period of time. There are a number of different mechanisms that may drive speciation. One of these is natural selection, which is a process that increases the frequency of advanta-geous gene variants, called alleles, in a population. Natural selection can result in organisms that are more likely to survive and reproduce and may eventually lead to speciation. A second process called genetic drift describes random fluctuations in allele frequencies in populations, which can eventually cause a population of organisms to be genetically distinct from its original population and result in the formation of a new species.[25]

The ancestors of modern primates split from the ancestors of modern rodents and lagomorphs (rabbits, hares and pikas) about 75 million years ago, the first human-like creatures appeared about 66 million years ago, and the first primates did not appear until around 50 to 55 million years ago.

Within the Hominoidea (apes) superfamily, the Hominidae family diverged from the Hylobatidae (gibbon) family some 15–20

---

[25] Speciation

million years ago; African great apes (subfamily Homininae) diverged from orangutans (Ponginae) about 14 million years ago; the Hominini tribe (humans, Australopithecines and other extinct biped genera, and chimpanzee) parted from the Gorillini tribe (gorillas) between 9 million years ago and 8 million years ago; and, in turn, the subtribes Hominina (humans and biped ancestors) and Panina (chimps) separated about 7.5 million years ago to 5.6 million years ago.[26]

The first human ancestors finally appeared between five million and seven million years ago, and modern humans originated in Africa within the past 200,000 years.

So, How Does Evolution Work?

Darwin once summed up natural selection in ten words: "[M]ultiply, vary, let the strongest live and the weakest die." Here "strongest," as he well knew, means not just the brawniest, but the best adapted to the environment, whether through camouflage, cleverness, or anything else that aids survival and reproduction. The word *fittest* (a coinage Darwin didn't make but did accept) is typically used in place of *strongest*, signifying this broader conception—an organism's "fitness" to the task of transmitting its genes to the next generation, within its particular environment.

---

[26] Evolution of Primates

"Fitness" is the thing that natural selection, in continually redesigning species, perpetually "seeks" to maximize.[27]

[Darwin's] theory is disarmingly simple. Darwin begins by noting the great competition in nature. Most species produce far more offspring than can survive.... However, most attempts at reproduction fail.... Sometimes, however, the local environment changes.... Such modifications to the environment pose new challenges.... What was useful in one environment is a disadvantage in another.... Natural selection tinkers with existing traits relevant to reproduction, making them ever more useful in the existing environment.[28]

And here is a more detailed explanation of **the four basic principles of evolution** mentioned above:

Variation in Populations
    In every species there is variation. This variability occurs even between related individuals. Siblings vary in color, height, weight and other characteristics. Other characteristics rarely vary, such as number of limbs or eyes.... Some populations show more variation than others, particularly in geographically isolated areas such as Australia, the Galapagos, Madagascar and so forth. Organisms in these areas may be related to those in other parts of the world. However, due to very specific conditions in their surroundings, these species evolve very

---

[27] Wright, R.
[28] Giberson, K. W.

distinct characteristics.

Inherited Traits

Each species has traits determined by inheritance. Inherited traits passed from parents to offspring determine the characteristics of the offspring. Inherited traits that improve the odds of survival are more likely to be passed on to subsequent generations. Of course, some characteristics, like weight and muscle mass, may also be affected by environmental factors such as food availability. But, characteristics developed through environmental influences will not be passed on to future generations. Only traits passed by genes will be inherited....

Offspring Compete

Most species produce more offspring each year than the environment can support. This high birth rate results in competition among the members of the species for the limited natural resources available. The struggle for resources determines the mortality rate within a species. Only the surviving individuals breed and pass on their genes to the next generation.

Survival of the Fittest

Some individuals survive the struggle for resources. These individuals reproduce, adding their genes to the succeeding generations. The traits that helped these organisms to survive will be passed on to their offspring. This process is known as "natural selection." Conditions in the environment result in the survival of individuals with specific traits which are passed through heredity to the next generation. Today we refer to this process as "survival of the fittest."[29]

Here's Where the Genes Come In

Your genes reside in you because they had a net positive effect averaged across all the individuals organisms and environments they have inhabited over thousands of generations.[30]

You probably have a general idea of how evolution—and genes—work, but I will first give a short review, emphasizing some issues that are especially important to the later discussion.

As you probably remember from your high school biology class, evolution works through genes. In humans, each cell normally contains 23 pairs of chromosomes, for a total of 46 chromosomes. Aligned along these "ribbons" or "strands" are chemical "letters" (genes).

You probably also know that computer language is "binary," which means that it uses only two "letters," yet the computer programs that have brought us all of the miracles of the "computer age" have been written based on this simple "two-letter" language.

Genes, on the other hand, have four chemical "letters" (C, G, T and A, which stand for cytosine, thymine, adenine, and guanine). These letters "spell" the "words" in our genetic "recipe" for making us who we are. These "letters"

---

[29] Masci, D. (1)
[30] Wilson, D. S.

always align themselves in "base pairs." (Adenine always pairs up with thymine, and guanine always pairs up with cytosine.)

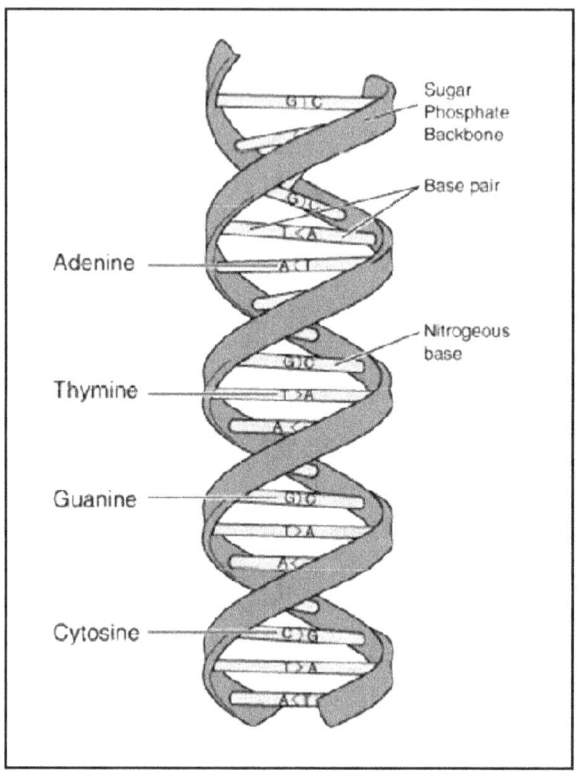

Our "recipe" is a long one. We humans have around three *billion* chemical base pairs ("letters") that make up approximately 20,000 to 25,000 genes ("words"). In each cell, our genetic code is "read" by other molecules which then make amino acids based on the "recipe." The amino acids are then, in turn, used to make the proteins of which we are made.

There is one exception to the rule that every cell in the body has 23 double-ribbon chromosomes: the cells (sperm cells in a man and egg cells in a woman) which create a new human being. When these particular cells are created, the body goes along the chromosomes and, like a modern-day shopper walking down a supermarket aisle picking items from the left- and right-hand "shelves," randomly picks genes from the two "ribbons" in each cell.

This means two things:

1) Only half of each parent's "recipe" is copied into the reproductive cells, and that half is combined with the other parent's "half recipe" of DNA to make up the full complement of DNA that is used to build the fetus.

2) With more than 20,000 "word pairs" to randomly pick from in each parent, every offspring is a completely unique individual with a set of genes that no other human has ever had before or will ever have in the future. (And remember that one percent of the time — about 200 times on average — a mistake is made that results in a "mutation."

Why don't we just divide into two one-celled microorganisms with the same genes as the parent? We don't because, if we did, all of our "children" would have the same DNA as we do. The goal of sexual reproduction is to continually "re-sculpt" populations of organisms so, generation after generation, they can continue to adapt to an ever-changing environment. (The environment you lived in yesterday was not the same environment you live in today, and the environment you will live in tomorrow will not be the same as the environment you live in today.)

Over many generations, the DNA of an organism is "sculpted" by its environment. For example, why do we humans have five fingers? If fingers are useful, why don't we have six or seven…or ten? We have five fingers because Nature, through empirical research on thousands of generations over millions of years, determined that while it is worthwhile to "invest" the body's resources in five fingers, it is not worth the effort to "invest" in any more than that. It costs the body energy to create, maintain and repair each of our fingers and Nature discovered that "budgeting" energy for five fingers made sense but that budgeting it for more than that did not.

So Why Hasn't Evolution Gotten It Right Yet?

You may be asking yourself, If evolution has been at work shaping us for hundreds of millions of years, shouldn't it have gotten us right by now? Why is the world (and, more specifically, human society) so imperfect? And why do so many people die of genetic diseases?

Well, there are several reasons we aren't all angelic little darlings. Among them are:

1) Mutations

Sometimes the body makes a mistake when copying the genes to make new cells (and this includes the reproductive cells). In fact, each of us has at least 100 new mutations in our DNA.[31] (However, as we will soon see, far from being a "mistake," in the larger scheme of things mutation is

---

[31] Chand, S.

necessary if a species is to adapt to the ever-changing environment.)

> An adaptation is anything brought about by the genes that helps them fulfill this metaphorical obsession, whether or not it also fulfills human aspirations. And this is a strikingly different conception from our everyday intuitions about what our faculties were designed for.[32]

2) Lag in adapting to ever-changing environments

The environment is always changing, and it always changes faster than our genes can catch up. This can be a lag of several decades or it can be a lag of thousands of years.

Our basic psychological and physiological makeup was formed over the hundreds of thousands of years our ancestors hunted on the plains of Africa, and it has still not "caught up" with the changes that have happened in our lifestyles over the past 50,000 years or so.

> [P]sychological mechanisms productive of adaptive behavior in a prehistoric past may, when operating in creatures no longer living in such an environment, result in different, surprising and/or maladaptive behavior.[33]

For example:

---

[32] Pinker, S.
[33] Miller, A.S. & Kanazawa, S.

> Researchers at Ben-Gurion University of the Negev (BGU) have discovered that gene mutations that once helped humans survive may increase the possibility for diseases, including cancer....
>
> "Our ancestors responded to environmental changes, such as climate shift, with mutations that increased their chances of survival. But today, these same mutations predispose us toward complex diseases such as cancer...."[34]

3) Selfishness

As Richard Dawkins has pointed out, genes are selfish little critters.

> According to Dawkins, our genes have programmed us to compete with other conspecifics for valuable resources. Those more successful at securing food, shelter, and reputation have a greater chance to reproduce successfully. And on the sociobiological account, we will attempt to perpetuate our genes any which way we can.[35]

All any gene "cares" about is doing what is necessary to maximize the odds that copies of it will survive into the future. This means that people (sometimes) act in self-centered ways:

---

[34] "Did Evolution Make Us Cancer Prone?" (
[35] Boyd, C. A.

Natural selection is the morally indifferent process in which the most effective replicators out-reproduce the alternatives and come to prevail in a population. The selected genes will therefore be the "selfish" ones, in Richard Dawkins's metaphor—more accurately, the megalomaniacal ones, those that make the most copies of themselves.[36]

However, as we will discuss later, it is often more effective (in the long run) for us to act in altruistic ways in the short run and our genes program us to do this.

4) Non-genetic causes: Abuse/ Poor Nutrition/Accidents/ Natural Changes in the Environment, etc.

5) As some members of a group become better adapted to an environment, more of them survive and have offspring and that creates greater competition for limited resources, which makes life harder for everyone, but especially for those who are less well adapted.

And finally, as we will discuss below:

6) Useful Mutations/Maladaptations: Nature discovered that it is important to have a supply of organisms which are *not well adapted* to the present environment, "waiting in the wings," so to speak, so they can potentially be called on in the future if the environment changes in a way that makes them better adapted.

---

[36] Pinker, S.

[N]o organism ever copies its genetic information perfectly, for if it did achieve such "perfection," it would quickly find itself at an evolutionary dead end. Unable to adapt to changing conditions or to new competition, before long, it would be driven to extinction by a host of more flexible competitors. As a result, evolution itself, driven by natural selection, favors organisms that are able to change.[37]

The Importance of Being Imperfect[38]

As mentioned above, in the larger scheme of things, mutation is necessary if a species is to adapt to ever-changing environments. To understand why, let's think (again) about the very first life forms to ever exist on earth.

"Life" started when, by random chance, some molecules were formed that made copies of themselves. However, they were able to do this in a specific chemical environment (at a specific temperature).

We can represent this in this way:

---

[37] Miller, K. R.

[38] Note: Much of the discussion in this section is based on the explanation in the book, *Nature and Man's Fate*, by Garrett Hardin. I learned more from that book than from any other book I have read about how life works. I highly recommend that you read (or at least a summary of its main points.)

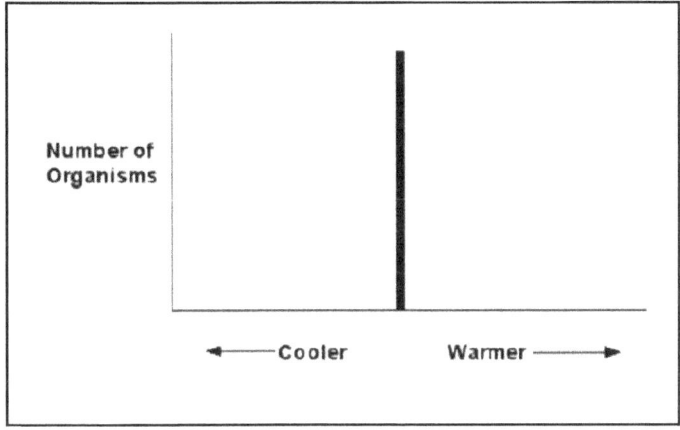

The figure represents a situation when all of the organisms are adapted to reproducing at a specific ambient temperature. As soon as they drifted out of this environment (or the environment changed) and the temperature changed, however, they (and all of the exact copies of themselves they had made) might lose the ability to replicate themselves.

So how could life continue? It could continue only if, in the process of replication, once in a while a "mistake" was made, i.e., an atom or molecule was *mis*-copied. But which atom or molecule should be mis-copied? And how often? And in which direction—to prepare (in the above example) for a warmer environment or a cooler environment?

The answer (and I want emphasize this) is that *no one—* not even Mother Nature (or God)—knows what will happen in the future! The reason why no one knows is that *no one can predict how the environment a species will have to survive in will change in the future.*

There is one thing that is almost always true, however, and that is that small changes are more likely than large ones. Therefore, it makes sense that Nature will "hedge her bets" by keeping most of the offspring of a species pretty much the same, changing some of them a little bit, changing even a fewer number more, and so on.

This can be represented in the following figure:

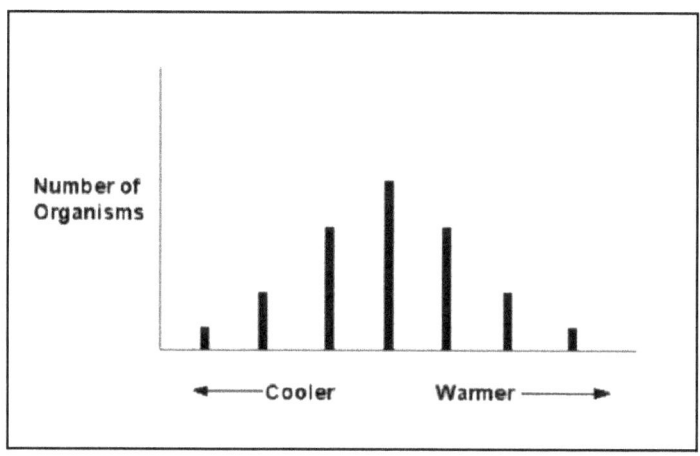

This means two things: 1) No matter how the environment changes (that is, whether it gets warmer or cooler, by a large amount or a smaller amount) some of the offspring will be well-adapted to that environment, which means that they will have a higher probability of surviving and reproducing and passing their characteristics on to the next generation, and

2) there will always be—and *must be*—a lot of organisms which are *not well adapted to the current environment*.

Remember: Mother Nature has had *billions of years* to work out the details of this plan, so by now she has got it down pretty good!

Now if you are at all familiar with statistics, the pattern in the figure above will immediately remind you of something called the "bell-shaped curve":

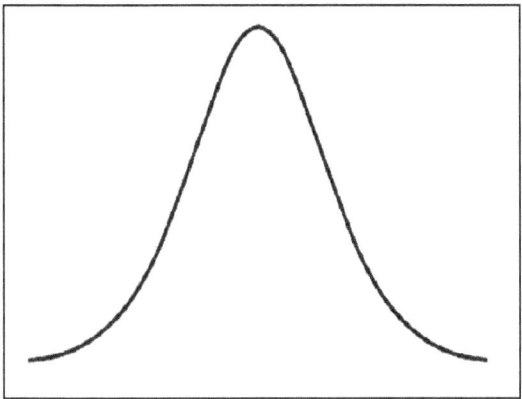

The "bell-shaped" curve is a general representation of the most common distribution of traits in any group of organisms, but its exact shape will vary depending on the circumstances: sometimes it's narrower, sometimes more spread out, sometimes lop-sided.

One more point needs to be made: In reality, the center of the bell-shaped curve is rarely at the point of "maximum adaptiveness" because, as the curve is shifting to the right or left, so is this environment. Therefore, the population of organisms is always "chasing" the point of maximum adaptiveness for every one of the thousands of traits it has. However, since the point is always moving, even the individuals in a population who are at the center of the bell-shaped curve may not be the best adapted.

To reiterate: no one—*not even Nature (or God)*— knows what will happen in the future! Distributing traits according to the bell-shaped curve is the *best way* to deal with the ever-changing nature of the environment *even if it means that some organisms are better adapted to survive in the current environment than other organisms are.*

In other words, over billions of years of experimentation, nature has determined that natural selection, with all its imperfections, offers the highest likelihood that genes will survive. Unfortunately, this means that many (maybe most) of the members of a species will spend their entire lives being less-than-perfectly adapted to their environment, and it is especially bad for those organisms on the "wrong side" of the curve when the environment changes in a direction favoring those on the other side of the curve.

The "Survival of the Fittest"

[A]ccording to Darwin himself, "survival of the fittest" and "natural selection" are basically the same thing—both phrases tell us that in any population, those individuals with characteristics well suited to the environment tend to be preserved, while those less well suited tend to die off.[39]

However, I do not think the phrase, "the survival of fittest," accurately describes the situation because the real situation is that *the fitter have a higher chance of surviving than the less fit,* but even an individual who is extremely "fit," genetically speaking, may not survive and pass on his genes, and one who is much less fit may do so.

---

[39] Smith, C. M. & Sullivan, C.

Thus, it's important to note that the reproduction of an organism at the center of the bell-shaped curve is not guaranteed and that an organism that is less well adapted is not doomed to not be able to pass on its genes. Rather, we can say that *there is a positive correlation between how well adapted an organism is and its chances for survival and reproduction*, which means that a well-adapted individual has *better odds* of survival than a poorly adapted one. However, the "better-adapted one" may not survive (it may be hit by a meteorite while walking to work) and the less-well-adapted one may survive (it may find a million-dollar lottery ticket while dumpster diving). In addition, an organism that is exceptionally well adapted in most of its traits may fail to survive and pass on its genes due a few (or even one) "killer" gene.

Of course, some traits (such as number and position of body parts) have been through the evolutionary beta testing process for so many years that virtually everyone shares the same traits, but there is considerable variation among the more than 20,000 other genes each human has.

This also means that no one is completely well adapted nor is anyone completely poorly adapted to their environment. We all carry an assortment of genes, from the very well adapted to the poorly adapted, and we as whole organisms compete to survive and reproduce against others who have similarly varied assortments of genes.

So how do we know who is well adapted?

Count the Descendants

Another way to judge how well people are adapted to their environment would be to look at who has the most descendants. (Remember that being "well adapted" means a relatively higher likelihood that an organism will survive and pass its genes on to descendants.)

For example, one might think that a person who has ten children is better adapted than a person who has few or no children. However, what if a person who has ten children has two grandchildren but a person who has two children has ten grandchildren?

So, the question is how many generations down the family tree should we look to judge how well adapted a person is.

The answer is, who knows? It is entirely arbitrary!

For example, if we had the genetic records of everyone who was alive in 1024 AD, and everyone alive since then, then we could determine which people alive 1000 years ago had genes which were most successful in surviving over time.

Likewise, if we could get the genetic information of everyone who is alive today and compare it to the genetic information of everyone alive 1000 years from now, we could determine whose genes today are best adapted to long-term survival.

Unfortunately, we don't have either of these kinds of information.

So, how can we identify which organisms (people) are "better adapted" than others?

This is a surprisingly hard question to answer because we have not yet developed means to judge this using either of the ways above (the number of a person's descendants to survive after an arbitrary number of generations/years) to judge how well adapted a person is to his environment. (And, unless we invent a way to see the future, it is unlikely that we will ever have this ability.) So *there is only one way to see whose genes are best adapted: sit back and let nature decide.*

The Grim Reaper Sculpts the Gene Pool

The environment is always changing. (The environment you lived in yesterday is not the same as the environment you are living in today, and the environment tomorrow will also be different than the environment you are living in today.) And by "environment" I mean not only the macro physical environment such as the climate, war and pestilence but also the micro-environment of pathogens and the social environment of everything, from which skills and talents are advantageous to earn a living at a particular place and time to what the opposite sex is looking for in a mate). Therefore, it is imperative that a group or a species adapt (change) if it is to survive.

> What Darwin called "natural selection" is nothing more than the sum of Nature's sorting processes. Random mutations that are *functional*, that help an organism survive or reproduce, will tend to be passed on to the next generation—not all

the time, but often enough to serve as a shaping force. Variations that are not functional tend to be deleted when the organisms that bear them falter in their ability to survive or reproduce. Functional mutations will be *inherited* by later generations; dysfunctional mutations will not. It is that simple.

Whatever the sources of *variation* in the genes…, it is Earth's climate, topography, chemistry and communities of life that put all novelties to the test. This is the sorting process of *natural selection*. Over eons, step by step, this natural sorting process has sculpted diversity and complexity in the stream of life….[40]

Unfortunately, nature is merciless, and there are (and must be) losers in the battle for survival. There are two things to remember in this respect.

The first is that, for a group of organisms to adapt to a changing environment, who dies and does not reproduce is just as important as who survives and reproduces. In general, the adaptation of a group can occur only when a larger percentage of the better adapted individuals survive than the percentage of less-well-adapted individuals. (And, as discussed above, we have no way of determining which individuals are "better adapted" other than waiting to see which ones survive and how many descendants they have. So, in a sense, the term "well adapted' is circular: by definition, "better adapted" people are those who have a higher chance of surviving and passing their genes on to

---

[40] Dowd, M.

their descendants, but we don't know which people they are until they do so!)

The second thing is that adaptation cannot take place unless more offspring are produced than can survive. The following figure may help explain this:

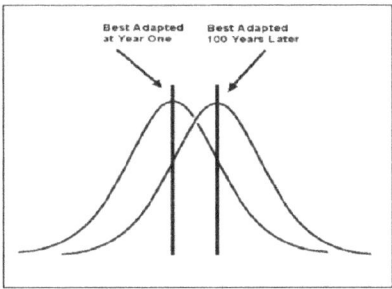

You can see that, over a period of time, the line which marks the degree of trait most valuable to survive has shifted to the right. This means that, if not interfered with, the entire bell-shaped curve will also shift to the right (so that the middle of the curve is closer to the *new* "optimally adapted" point) in order to prepare for changes in the future (which may require that the curve shift either back to the left or further to right.) But the only way that Nature can "sculpt" a new curve in each generation is if she has a large amount of "raw material" to start with.

In every generation, people on all parts the continuum will have children.

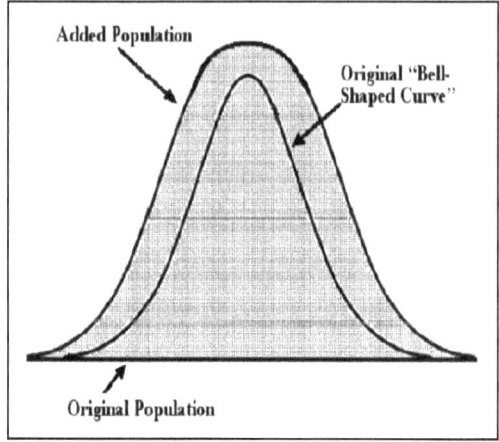

Referring to the previous figure, as *the distributions of traits in a group* "moves to the right" to adapt to the new environment, *fewer* people who are on the left-hand side will survive (or, in other words, more will die without passing on their genes or they will have fewer offspring who survive) and *more* people on the right-hand side will survive.

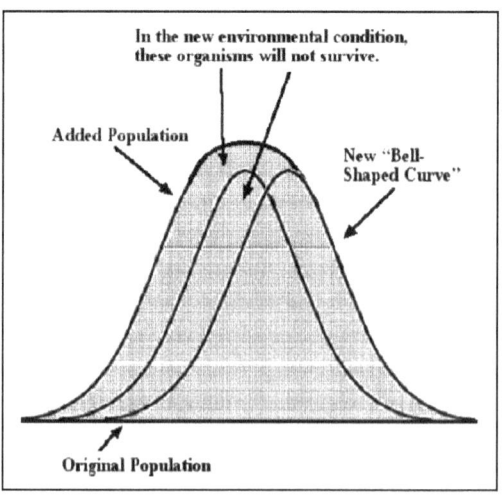

Here it is once again important to point out that no one—not even Mother Nature—knows in which direction the "well-adapted" line will shift in the future, so in order for her to do her work it is important that *all of the groups of people on both sides of the line have more children than can survive, that not all of the children survive and have offspring,* and (as explained above) *as few artificial (external) means be used as possible to help less-well-adapted individuals survive.* (See Part Two for more on this issue.)

> Evolutionary change comes about at the level of populations, not at the level of individuals.
>
> As circumstances change, the combinations of genes that produce the most successful phenotype [the observable properties of an organism that are produced by the interaction of the genotype and the environment] also change. Evolution is the gradual adaptation to new environments by selection of gene combinations that give the greatest reproductive success.
>
> But individual organisms cannot change their combination of genes. They are stuck throughout their lives with the combination of genes donated to them by their parents.
>
> During the process of sexual reproduction, these combinations of genes are shuffled and redistributed every generation. Among many progeny, therefore, it is likely that some of them will have a combination of genes that allow them to survive in a changed environment.

> These offspring will survive, enjoy reproductive success, and pass on more of their genes than those individuals that are less well adapted. Subsequent generations will possess more and more genes for survival in the new environment than did their ancestors. Slowly, the population will become more fully adapted to its new environment.[41]

So what does "well adapted" mean?

There are two ways to judge well an organism is "adapted" to its environment.

The First Way: Organisms which are better adapted are able to survive "more cheaply" (i.e., are less dependent on outside resources).

Organisms use two means to survive in their environment, internal means and external means, and those who are born with more internal resources and less need to depend on outside resources have a better chance of surviving.

* * * * * * * * * * * * * * * * * *

Internal Means

"Internal means" refers to ways that do not depend directly on anything outside the organism. These include a) the organism's genes, and
b) the organism's normal bodily processes/ functions. These two are relatively "cheap" in

---

[41] "The World of Darwin: The Passage of Time"

that the organism does not have to expend much energy to
make them happen: the organism is
(in most cases) born with the necessary genes
and the bodily processes (such as digesting food
and fighting invaders such as viruses) occur automatically
and do not require a lot of energy.

*The meanings of "cheap" and "expensive" in this
context*

Note that in this context the words "cheap" and
"expensive" refer to how much energy an organism needs
to expend ("spend") in energy to accomplish something.
This is related to the normal uses of the word: Something
that is "expensive" in the world requires more energy to
attain or take possession of than something that is "cheap."

Humans have invented a social way to save
the energy we expend and to measure how much is
necessary to attain something. That way is money and
prices. Yes, money is a way of energy saving! Instead of
receiving an immediate benefit (such as food) for their
effort, people receive a token (money) which they can save
and use to buy things later on. People who have worked
more than others (in terms of efficiency of their work and
time they have spent working) receive more money for
their work, and people who expend less energy or work less
efficiently receive less money.

A Get-Rich (but not quickly) Scheme

Of course, some will point out that ditch diggers who
work 50 hours a week expend more energy and get paid
less than, say, a designated hitter in baseball who "works"
only a few hours a week and gets paid millions a dollars a

45

year. There are two reasons this situation exists. One is that different people are born (because of their genes) with different levels of adaptiveness to their environment. For example, some people are born with innate potential to become great athletes or musicians or engage in occupations that require great mental ability (such as law or medicine) or physical ability (to hit a curveball or to slam dunk, for example). (Believe me, *no one* in *my* family has been born with any musical or artistic ability!) People born with great innate ability are rare, so they can get paid more for their expenditure of energy because they can get more done (in terms of value to others) than those who are born with less innate ability. (There is one exception to this rule: people who are born with fewer innate abilities can still become rich by becoming politicians. ☺) This is especially true if they are especially adept at doing things that few people can do.

However, there is another way (besides finding rich parents to be born to) to increase one's ability to earn more money with less effort: increasing one's skills. This can be done by: 1) studying (acquiring knowledge), 2) practicing (acquiring skills), and 3) acquiring an ability (through study and practice) that is rarer and thus in more demand.

However, as any adult who is still paying off their college student loan can attest, studying usually requires a lot of money, whether it is paid by the student himself (or his parents) or by society. So why do some people study hard? It is because it is assumed that their immediate expenditure of time, money and effort will be more than repaid in the future when they are able to get a more high-paying job.

Practice also requires time, effort and (usually) money. In his book, *Outliers*, Malcolm Gladwell notes that becoming world-class in something usually requires about 10,000 hours of dedicated practice, which works out to about three hours a day for ten years. Again, some people are willing (and able) to expend their time, money and energy in practicing this much because they foresee that they will be more than repaid for their efforts in the future.

A third way is to acquire an ability that (because of the law of "supply and demand") is much needed but hard to find in society, such as a master craftsman. But there are reasons that few people have these skills — they may require a lot of study and practice to acquire, or the job might be less rewarding or more dangerous (or might not even exist by the time one becomes an adult!).

External Means

"External means" refers to what the organism cannot make internally so it needs to be gotten from its environment. Humans, for example, need oxygen (which is relatively "cheap"), food (more expensive) and (at times) medical care (potentially very expensive).

As technology advances, life is getting easier and easier as more and more people rely more and more on external things rather than what they were born with. However, this also causes a problem.

As mentioned above, things controlled by our genes are relatively "cheap" — in fact, our genes are "free." However, things outside of us that we depend on to survive are (relatively) expensive.

Each organism has a "budget" of *internal resources* (abilities produced internally and energy) and *external resources* necessary to get, do or control the things needed to survive. (Some environments are more "expensive" for an organism than others.) Those who are more efficient (need to "spend" less time, energy and money to acquire more resources [money] without expensive outside help) are more likely to survive and pass on their genes to the next generation, so we can say that they are "better adapted" to their environment. However, the more an organism depends on things outside itself in the environment to survive, the more difficult it will be for it to survive, so it is less well adapted. In addition, internal resources are more dependably available than external resources.

So, in addition to *savings* one has to accumulate and the amount (and efficiency) of one's *efforts* to acquire external resources, the *efficiency* and *rarity* of one's abilities are also important.

So, one way to judge how well adapted an organism is to measure its lack of dependence on outside sources, and its ability to produce more energy (money) that it needs to stay alive. This is shown in the following equation:

> **Internal Resources + External Resources**
> **divided by**
> **"Cost of Living"**
> **(Amount of Energy Needed**
> **to Survive and Reproduce)**
> **equals:**
> **Likelihood of Survival**

One Last Point

One last point needs to be emphasized: Just because someone is not "well adapted" to live in a particular environment does not mean that he or she should be banished from the earth! For one thing, someone who is well adapted to one environment may not be so well adapted to another environment. (If you kidnapped me and dropped me off in the middle of the Amazon rain forest, for example, I would not be "well adapted" to surviving there!) So, someone who is not well adapted to one environment can simply move to another environment.

Second, as discussed above, an individual can improve his or her skills (and lose bad habits) so he or she becomes better adapted to his or her environment.

Third, while one may not be "well-adapted" in respect to one trait, he may have other traits which more than make up for his lack of adaptiveness on another trait.

In addition, as explained above, nature needs to keep some less-than-well-adapted organisms around for when the environment changes, and this is why "eugenics" (which will be discussed in the next section) can't work: First of all, (as explained above) no human can reliably define what "well-adapted" or "mal-adapted" means, and, secondly, we need "mal-adapted" people to be kept "in reserve" in case the environment changes in "their direction."

Last, if someone has not had any children, it does mean that their life is wasted or meaningless, because they can help the survival of other people with whom they are most closely related genetically, such as close relatives.

From the above discussion, you might get the idea that we are doomed to live a life of "dog-eat-dog" competition. However, as Peter A. Corning (Director, Institute for the Study of Complex Systems) (as quoted by Dowd, p. 35) writes:

> Ecological communities are not simply gladiator fields dominated by deadly competition; they are networks of complex interactions, of interdependent self-interests that require mutual adjustment and accommodation with respect to both the other co-inhabitants and the dynamics of the local ecosystem. The necessity for competition is only half of a duality, the other half of which includes many opportunities for mutually beneficial co-operation.

We will have more to say about the importance of altruism and cooperation in the section where we discuss the question, "Is Evolution antithetical to morality?"

Part Two:

What can we learn from Evolution?
Implications and Ramifications[42]

---

[42] Note: If you are a leftist/socialist/liberal, you may want to take a sedative (or smoke a joint) before reading this section. ☺

Now we get to the tough part! In this part I am going to use what we know about evolution, natural selection and adaptation to discuss and examine various social policies.

Warning: Some of the things I point out will no doubt surprise, enrage or even disgust you! But remember: I am not advocating for or against any political policies or social mores.

I will just objectively explain how evolution works, based on evidence and logic. It is up to us to decide what to do with that knowledge.

First of all, what is "life"?

You probably think that is a silly question! The answer is obvious — all living things (plants and animals). But Mother Nature looks at it in another way.

To nature, *life* is everlasting. In fact, that is her primary goal: to keep the whole life show going for as long as possible.

But do individual plants and animals live forever? Of course not. But what does?

I might answer by saying "DNA," but the individual molecules that make up DNA are not eternal. They serve as a template to make new copies of the genetic code, but they themselves move on to other tasks when their host dies.

So what does! *The genetic code!*

Of course, the genetic code is not immutable — it changes with every generation — but the parts of it that help its organism survive and reproduce continue to "live."

So what was Mother Nature's secret? (OK, here comes the big secret that Mother Nature doesn't want me to reveal because it is surprisingly simple.)

Actually, all Mother Nature did was to start the cascade of life by creating *one molecule* that could replicate itself. That's all! It was like kicking a stone at the top of a mountain that created an avalanche.

But that act was a work of genius!

You see (as explained above), once that one molecule started reproducing itself (with occasional mutations thrown in), Mother Nature *didn't have to do anything* because everything that happened after that came (if you will excuse the expression) "naturally." As new versions of the molecule were produced, ones that were better adapted to surviving and reproducing in their environment would create a new generation of themselves.

After that, it was a simple matter of combining with other molecules, making the genetic code longer, more complicated, and more diverse. This, in turn, led to the development of the amazing number of different life forms we see today.

"But wait!" I hear you thinking. "Are you saying that plants, animals, humans, bacteria, mosquitoes, etc., are not alive?"

Well, they are "alive" in the sense that they are created, grow and exist for a while, but that is just a consequence of Mother Nature's original act.

You see, *all living things are just that —"things" — that DNA creates, using its own DNA code, as a way of "implanting" parts of its own code into a new "thing"(or things) that will continue the process.*

Now, if you can understand and accept that, it puts "life" in a whole new perspective.

And it also raises another important question: "What is necessary to keep 'life' (i.e., the genetic code) alive?", and the answer is a surprising one: Just a few basic things that are needed to keep DNA's "hosts" alive long enough to pass on the code:

1) A Genetic Code that is well enough adapted to the organism's environment that the organism survives and reproduces.

2) Shelter: protection against the elements and dangerous threats

3) Energy. This is provided by the intake of various necessary elements and molecules (commonly known as, "eating food and drinking water").

4) Protective Covering, such as skin, clothing, fur, cell walls, hair, etc.

5) Access to Resources (such as water, land, nutrients, raw materials, etc.) that are necessary for the organism to stay alive.

6) The ability to produce more offspring than can possibly survive in a given environment so the group/species as a whole can adapt to as their environment changes.

The above is what is necessary for 99.99% of organisms to survive. However, one species needs not only the above, but three other things:

7) An environment that allows them to survive.

8) A record of things that previous generations have found to be useful in helping them survive (in other words, a culture, lore, history) that are not passed on by the genes.

And one last thing that is perhaps the most important of all:

9) *A desire to survive*, reproduce and to protect the previous two things (the environment and the records of ideas that existed before they came into the world).

Of course, that species is *humans*.

Next, I am going to explain how, in the pursuit of "progress," humans in modern Western societies have made their survival *less likely* in the long term. I call this:

Anti-Naturalism[43]

"You are about to enter another dimension.
A dimension not only of sight and sound but of mind.
A journey into the wondrous world of science.
Next stop— The Anti-Natural Zone!"

I coined the term "Anti-Naturalism" to refer to attempts by humans to ignore or contravene the laws of Nature in order to make their lives easier and to live longer.

Birth Rates

What has happened since the advent of Women's Lib is that women (and especially talented women) are more likely to devote their efforts to "reaching their full potential" than to having and raising children.

---

[43] I chose the term "Anti-Naturalism" because "unnatural" has some connotations that are unrelated or irrelevant to what I want to discuss.

What does this mean for the future of society?

In the short run, it is good that society is finally benefiting from all of this heretofore unused feminine talent. However, at the same time, the better adapted a person is, the more important it is to *society* that he or she have children so as to contribute his or her genes to the future gene pool of the society.

However, according to Wikipedia:

> There is generally an inverse correlation between income and the total fertility rate within and between nations. The higher the degree of education and GDP per capita of a human population, subpopulation or social stratum, the fewer children are born in any developed country.[44]

It seems safe to assume that there is (on average) a direct correlation between what a woman earns, how talented she is, what she contributes to society through her work and the value of her labor. And isn't it also safe to assume the children of more talented women will (because they inherit half of their genes from their mother) also eventually contribute more to society than the children of less talented women?

Unfortunately, the current situation is that (on average) the *less a woman makes* the *more children* she will have. Here is the latest data from statista.com:

---

[44] "Income and fertility"

**Birth rate [per 1000 households]
in the United States in 2019[45], by household income[46]**

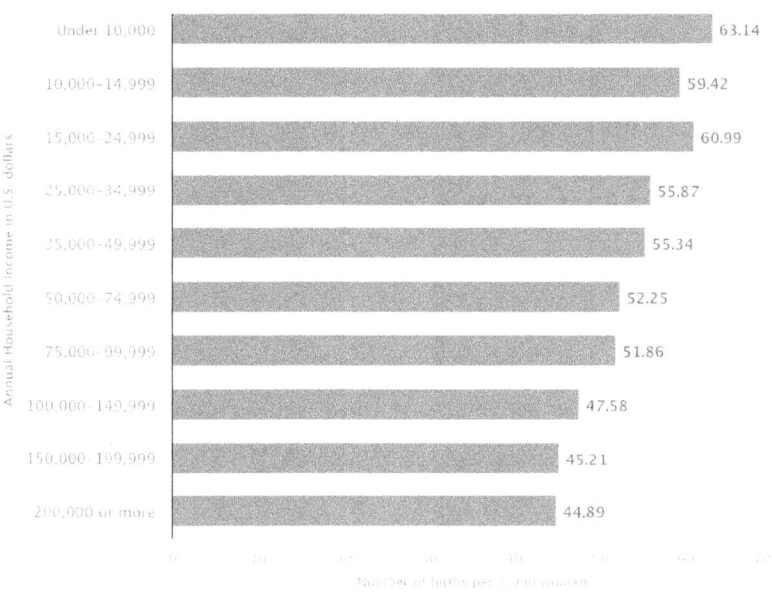

iii

According to the above, women in families in the lowest income brackets have higher birthrates than women in the highest income brackets, and, as noted above (at least in Bolivia), there was "an inverse relationship between social class (and the educational level) and fertility."

Of course, it has long been recognized that women in poor countries tend to have more children than women in richer countries for two reasons: 1) Poorer countries have higher levels of infant mortality, and 2) women (and their husbands) in the poorer countries cannot depend on their

---

[45] Unfortunately, this is the latest data available.

[46] "Birth rate in the United States in 2019, by household income."

savings or a government "safety net" to take care of them in their old age so having a lot of children is a form of "social security." (By the way, this is another example of how socialism is bad for our genes: people who don't believe they will need children to take care of them in their old age have fewer children, which means fewer chances to pass on their (better adapted?] genes plus fewer workers to contribute to the "social security" of the elderly.)

Finally:

A women's educational level is the best predictor of how many children she will have, according to a new study from the National Center for Health Statistics, Centers for Disease Control and Prevention. The study, based on an analysis of 1994 birth certificates, found a direct relationship between years of education and birth rates, with the highest birth rates among women with the lowest educational attainment.[47]

So, according to the above, the higher the social class, the higher the income, and the higher the education level a woman has, the fewer children she is likely to have.

What would you say if a foreign enemy came up with a drug that lowered the birthrate of the very people whose genes can contribute the most to the future of society? You would say that it is an outrage, an act of war, even a crime against humanity! Yet that is what "women's lib" has done! What has happened over the past 40 years or so is that the birth rate of the "best and brightest" women in western countries has fallen.

---

[47] "Mother's Educational Level Influences Birth Rate"

In an ideal society, those who are best adapted to the environment would have the highest birthrates, but the opposite is happening in Western countries today, and what are the long-term consequences for Western society if the most talented women are less likely to pass their genes on to the next generation than less talented women are? And why should the federal government encourage this trend through its "income redistribution" policy?

The fact is that through their "anti-eugenics" policies of taxing the rich in order to subsidize the poor, liberals make it harder for talented women to pass their genes on to the next generation while making it easier for less well adapted women to do so.

Now, I know what you are thinking: Women are not baby-making machines!

That's true. However…

(Now buckle your seatbelt—this may get you a little excited….)

One must ask the question: In which way does a highly talented woman contribute more to society over time—by using her talents to contribute to society over the forty to fifty years of her working life at the expense of passing fewer of her genes on to the next generation or by "wasting" her talents while passing her genes on in her children?

There are too many examples of contemporary women who have sacrificed childbearing for their careers, but let's discuss one: Hillary Clinton.

Everyone would agree that Hillary Clinton is an incredibly talented woman who has contributed a tremendous amount to society in a variety of roles.

However....

If, instead of her making those contributions, a slightly less talented man had held her positions and she had had three or four or more children, wouldn't she have contributed more (by passing her talented genes down to her descendants) over the future hundreds or thousands of years than she did in her career?

Of course, in terms of society, this means that: 1) the average level of talent of those contributing to society right now would fall (because many positions will be filled by less-talented men than the women who could have held them), and 2) many women will never reach their "full potential" and will instead stay home raising their children.

Can you say, "1950's"?

The fact is (uh, oh—here it comes) that, from our genes' point of view, the best strategy to ensure the future of society (and their own future) is for there to be a direct correlation between women's adaptiveness —that is, how well they are adapted to the environment—and the number of children they have. (Remember: Genes don't care a whit for our feelings or our emotional needs.) Unfortunately, now the exact opposite is happening.

Now I know you are thinking, "That's ridiculous. A lot of talented women have children *and* a career."

That's true, but, over time, the number of people in any group will either rise or fall. And if the result of "women's liberation" is that the average birth rate of talented women has fallen even a small amount from *above* the replacement birth rate of 2.33 (see below) to *below* it, the proportion of talented women in society will also decline along with their contribution to the future society.

And now:

Women in the richer countries average only about 1.7 children each during their lifetime, while those in the poorer countries average 3.6.[48]

So, over time, there is a big differences between a birth rate in a richer country of 1.7 children per woman versus 3.6 children per woman in poorer country: The population of an poorer country whose women have, on average, 3.6 children each, will continue to grow (and double every 20 years!) while a richer country whose women have, on average, 1.7 children each, will disappear from the face of the earth!

Birth Rates by Location in 2024 (Estimated)

By Continent

World	2.31
Africa	4.12
Asia	1.93
Europe	1.51
Latin America and Caribbean	1.83
North America	1.65 [49]

---

[48] "Third World population growth at record high."
[49] "Fertility rate in each continent and worldwide, from 1950 to 2024"

By Country

So which countries have highest and lowest birth rates?

In 2024, there are six countries, all in Sub-Saharan Africa, where the average woman of childbearing age can expect to have around six or more children throughout their lifetime. In fact, of the 20 countries in the world with the highest fertility rates, Afghanistan is the only country not found in Sub-Saharan Africa.[50]

And here is a list of the ten countries with the lowest birth rates (estimated) (with their birth rates in parentheses) in 2024:[51] Taiwan (1.11), South Korea (1.12), Singapore (1.17), Ukraine (1.22), Hong Kong(1.24), Macau(1.24), Moldova (1.26), Puerto Rico (1.26), Italy (1.26), Spain (1.3),

Plus China has a birth rate of 1.7 and India's is about 2.0-2.1

Based on the above, I think it is fair to say that less developed countries have higher birth rates than more developed countries.

Now liberals in more developed countries will applaud the fact that, with the help of technology, they have managed to lower the birth rates in their countries. But what is the point of lowering birth rates to the point that those groups of people will become extinct?

Dear reader, are you afraid of "Global Warming"? Are you afraid of the how many people might be killed by a worldwide flu pandemic? Well, read the list of countries

---

[50] "Countries with the highest fertility rates 2024"
[51] "List of countries by total fertility rate"

above who have fertility rates *below* the average replacement fertility rate of 2.3 births per woman and realize that it is a list of places whose inhabitants are *on death row*!

If the present trend of low birth rates continues, all of the above nationalities will someday be groups of people that future humans read about only in history books!

(Of course, many of these countries accept large number of immigrants from nations with expanding populations in order to fill all the menial jobs, and, as a result, these Western nations will continue to have increasing populations.)

You may say, "Isn't it better to have a small population with high living standards than a larger one with lower living standards?" This is the opinion of some people, such as (liberal) Nobel Prize-winning columnist Paul Krugman of *The New York Times*, who believes that Europe isn't "the stagnant, decaying economy of legend."

> [T]he story you hear all the time—of a stagnant economy in which high taxes and generous social benefits have undermined incentives, stalling growth and innovation—bears little resemblance to the surprisingly positive facts. The real lesson from Europe is actually the opposite of what conservatives claim: Europe is an economic success, and that success shows that social democracy works.[52]

---

[52] Krugman, P.

He believes that Europe demonstrates that "social justice and progress can go hand in hand."[53] However, like many liberals, Krugman judges societies mainly in terms of their material success, ignoring their moral health and their long-term chances of survival.

On the contrary, Lord Sacks, Chief Rabbi of the United Hebrew Congregations of the Commonwealth since 1991 and the leader of Britain's Jewish community, claims that Europe's "population is … in decline, compared with every other part of the world, because non-believers lack shared values of family and community that religions have."

> Lord Sacks said: "Parenthood involves massive sacrifice of money, attention, time and emotional energy.
> "Where today in European culture with its consumerism and instant gratification – because you're worth it – where will you find space for the concept of sacrifice for the sake of generations not yet born?
> "Europe, at least the indigenous population of Europe, is dying."
> "That is one of the unsayable truths of our time…."
> "It may not be religion that is dying, it may be liberal democratic Europe that is in danger, demographically and in its ability to defend its own values."[54]

---

53 Krugman, P.
54 "Europeans too selfish to have children, says Chief Rabbi."

The fact is that, like the frog who is put in a pot of water which is slowly heated and thus doesn't realize the need to jump out until it is too late to do so, Europe society is comfortably dying and, if current trends continue, its culture will someday be supplanted by Middle Eastern or African culture.

ZPG vs. Expanding Population

Ever since the publication of Paul R. Ehrlich's best-selling book, *The Population Bomb*, in 1968, "zero population growth" has been one of the goals of the liberal agenda. The book predicted disaster if population growth was left unchecked and claimed that radical action was needed to limit the overpopulation.

The problem is that "The Theory of Evolution" tells us that (at least in the long term) societies cannot survive *without* population growth (or, to be more precise, giving birth at greater than the replacement rate). This is because (as explained in Part One above) any society which has "zero population growth" cannot adapt to the ever-changing environment.

To understand why, let's us an analogy:

Let's compare the situation at two factories in two different cities. Both factories employ 10,000 people.

The first city has "zero population growth," which means that decade after decade women have only enough children to keep the population constant.

Unfortunately, over time, the needs of the factory change. Some of its products no longer sell well, so it has to shut down old product lines and bring in new machines and new technologies to produce new products, so it needs people with new skills to run those machines. However, year after year it has basically the same labor pool to choose from. Whenever 100 people retire, about 100 young people show up to apply for their jobs.

The result will be that everyone in the city who wants a job will be able to get one but, over time, the output of the factory will decline because it has such a limited pool of labor to choose from in hiring new workers and, since the workers will essentially be guaranteed a job (that is, there is no competition for survival), there will have no incentive to improve their skills to meet the requirements of the changing jobs they will be doing.

The end result is that, as the products produced by the factory decline in quality, the sales of the factory decline and the standard of living of everyone who lives in the city declines.

The second city, on the other hand, has robust population growth. This means that every time 100 people retire from the factory, 150 people show up to apply for their jobs. This allows the factory to choose among them and to hire the 100 who are the best "fit" for the jobs.

This is how nature works. She needs a broad array of "applicants" for the future jobs of surviving in the ever-changing environment. She "selects" those who are best fit to survive.

But why doesn't the first city institute a program to train its young people for future jobs at the factory?

The answer is that no one knows what talents and abilities those future jobs will require, and it is the same with nature: no one, not even Mother Nature, can predict how the environment will change and, therefore, which traits will be necessary for the survival of future generations. So the only way to make sure that an adequate "work force" will be available is for there to be a large "labor pool" with a *random* assortment of traits (distributed, of course, according to the bell-shaped curve) from which to choose.

Now a liberal might prefer to live in the first city because, unlike in the second city where there will be high unemployment and some people may even die of starvation, in the short term at least there will be full employment in the first city and no one will "suffer."

The problem is that, in the long term, the first city will become weaker and weaker economically and will fall apart. In the second city, there will be "winners" and "losers" but those who survive will thrive. And eventually the first city will cease to exist or will be conquered by the "strong" city.

There are also economic consequences to zero population growth (or a declining population). One of the places where this is being seen is in Japan, where:

> The average number of children a woman gives birth to in her lifetime fell to 1.26 in 2022 from 1.30 a year earlier, tying the record low from 2005, according to the annual population statistics. The fertility rate is far below the rate of 2.06 -2.07 considered to be needed to maintain a population.
>
> Japan's population of more than 125 million has been declining for 16 years and is projected to fall to 87 million by 2070. A shrinking and aging population has huge implications for the economy and for national

security as Japan fortifies its military to counter China's increasingly assertive territorial ambitions.[55]

And what happened when China tried to limit the number of children people could have?

Over time, though, the policy's effect on the population's fear of the government and perception of babies meant that abortion and infanticide became a way of life. China's infanticide epidemic was particularly brutal on female children. Since male children were both socially and economically more valuable, especially in rural areas where they were needed to both work and inherit family farms, it was common for parents to kill or abandon female babies in favor of trying for sons.

It's become clear that the One Child Policy has backfired in more ways than one. The rapid decrease in population means that China's population pyramid is "top heavy," with older generations, especially the 45 to 65-year-old brackets, outnumbering younger ones. As older generations begin exiting the workforce, it's unclear whether younger generations will be large enough to care for them, both in terms of the number of healthcare workers and the direct care parents typically rely on from children, but also the economy and tax pool more broadly.

The gendered consequences of the one-child policy are similarly problematic. As a result of decades of sex-specific abortion and infanticide, the gender imbalance in the Chinese population is severe, especially for younger

---

[55] Yamaguchi, Y.

generations. With almost 35 million more men than women, many men are finding it difficult to find a wife, never mind having children....

Ultimately though, the biggest consequences of the one-child policy aren't demographic, they're value-based. Generations of small families, abortion, and infanticide have reduced the value many young Chinese citizens place on babies. Being raised around only children also has made it nearly impossible for young citizens to picture what life looks like with bigger families, and many don't see the point in having children at all. Those who do would rather spend their time and money on an only child, optimizing for the best schools, care, and resources for a single child.[56]

Another problem with ZPG is, who is going to pay off the debts run up by the previous population's spending on entitlements (not to mention, pension plans)?

Economic growth won't last as the U.S. labors under the burden of growing entitlement programs and weakness around the world, former Federal Reserve Chairman Alan Greenspan told CNBC.

The long-time central bank chief repeated his warnings about the weight that Social Security, Medicare and other programs are having on what have been otherwise solid gains over the past few years.

"I think the real problem is over the long run, we've got this significant continued drain coming from entitlements, which are basically draining capital investment dollar for

---

[56] Clough, A.

dollar," he told CNBC's Sara Eisen during a "Squawk on the Street "interview.

"Without any major change in entitlements, entitlements are going to rise. Why? Because the population is aging. There's no way to reverse that, and the politics of it are awful, as you well know," Greenspan added.[57]

Summary: ZPG vs. Expanding Population

How can Mother Nature ensure that more of the well-adapted people survive if there is zero population growth and how can she insure that fewer of the less-well-adapted survive if they are helped to survive artificially? In the long run, *nobody* will survive unless women have more children than can survive. Or, to put it another way, if 99% of a population have lower-than-replacement-level birth rates or practice "zero population growth" and one percent have higher-than-replacement-level birth rates, eventually the 99% of the population will cease to exist and the entire surviving population will be descendants of the one percent.

Remember that, in order to survive, all species must produce more offspring than can possibly survive and Nature decides which ones "make the cut." In every generation Nature needs a large block of marble to start with so, with each new generation, she can discard what she doesn't need and save what she does need.

I know this all sounds so unfair, so mean, so, so ... insensitive. Sadly, nature is (in Tennyson's words) "red in tooth and claw" and, in the end, merciless to those who do not have the traits to survive.

---

[57] Cox, J.

As explained above, from our genes' point of view, the key to success is to have lots of children, even if some of them will not survive and reproduce. Similarly, some religions (such as Islam, Catholicism and Mormonism) discourage their adherents from using birth control while at the same time encouraging them to have lots of children, but only within marriage.

Which is closer to nature's plan—to "control" or "repress" our natural instincts and limit the number of children we have by artificial (unnatural) means or to *let nature decide* how many children we have and then, later, after they have been born and have been given the chance to compete in life, let nature decide how successful they will be in surviving and reproducing?

This is not to say that it is always better to choose "quantity" of life over quality of life for our offspring and for the world. But, as explained above, a society which practices "zero population growth" cannot survive in the long run because it will not be able to adapt to changing environmental conditions.

Look at the world today. Which countries have the highest birthrates—the scientifically advanced countries or the so-called "developing" countries? Which people will be more likely to survive (and pass on their genes) when the next mega-catastrophe strikes the earth—those who have low birthrate and are increasingly dependent on all sorts of governmental programs and scientific and technological systems or those who don't have the luxury of a high standard of living so must depend on the biological abilities and defenses they were born with? Maybe it is true that "the meek shall inherit the earth."

But what about overpopulation?

An expanding population (that is, when the birthrate is above the replacement rate) is natural for three reasons.

First, as explained above, in order for a species to survive, the species has to adapt to an ever-changing environment, and the only way it can do this is to provide more "raw material" for Nature to use than can possibly survive.

Second, as technology improves, the number of people who can survive in a particular environment increases. And who has the moral right to say how many is too many?

Third, various places on earth, and the earth itself, may someday experience a catastrophe (such as a meteor strike or a pandemic) that kills large numbers of people. How many people will be necessary so that some of the local or global population will survive? Nobody knows! So, once again, it is best to let Nature decide because she has a lot more experience with this type of thing than we do!

Sex

Liberals generally have a more permissive attitude toward sex than conservatives do. The idea of "repressing your feelings" is anathema to them. After the development of the birth control pill (not to mention *Playboy* magazine) in the 1950s, the so-called "sexual revolution" in the 1960s, and the legalization of abortion in the 1970s, what used to be called "fornication" (consensual sexual intercourse between two persons not married to each other) became socially accepted. So, now that out-of-wedlock birth is more socially acceptable, there is less social pressure to abstain from sex before marriage or to marry a woman after getting her pregnant.

The result? In 2022 an unbelievable 40% of all babies born in the United States (and an incredible 70% of black babies!) were born to unmarried women![58]

What does this have to do with our genes? It is obvious that a baby born to a mother who has a partner committed to helping support and raise the baby will have a better chance of growing up to be a person who can live independently and pass his or her genes onto the next generation without government help. In addition, numerous studies have found that children born to single mothers face all sorts of disadvantages that children of two-parent families do not face.

Unfortunately, many feminists have insisted that men and women are essentially the same and that there is nothing wrong with women acting the way men have always acted. However, in genetic terms, it is advantageous for men to have sex with as many women as possible as a strategy for passing on as many copies of their genes as possible. Women, on the other hand, can have fewer offspring than men, so, in terms of passing their genes on, it is better for women to be much more selective in whom they have sex with.

Birth Control

Isn't it a bit ironic that the very people who are always talking about "nature" are the very ones who most strongly endorse the unnatural practice of birth control? If liberals so

---

[58] "Births: Final Data"

believe in nature, why don't they trust "nature" to decide who gets pregnant, when they get pregnant and how many times they get pregnant? (If nature learned that the ability to prevent getting pregnant or to abort a baby was good for the genes, then why didn't she give women these abilities?)

From society's point of view, there are two problems with birth control. The first is that irresponsible woman (or teenage girls who are too young to take full responsibility for their sexual behavior) are less likely to use birth control than more responsible women are. Over time, this increases the proportion of babies born in the society to irresponsible or less-responsible females. In other words, it helps relatively more "irresponsible" genes to be passed on to the next generation.

The other problem with birth control is that it may make the total birthrate go down to below replacement levels. (See the discussions on low birth rates above.)

Now here is a question that is either going to make you "think a second time" or make you start throwing burning bras at me:

What if the human race has developed in such a way that *any society in which the women have the ability to control how many children they have cannot survive because not enough babies will be born?*

Just asking….

Abortion

Abortion is another "unnatural practice." (If the ability to abort one's baby was beneficial, then why didn't that ability evolve in women?) Liberals see abortion as a fundamental right a woman should have, but, from our genes' point of view, there is no difference between killing a fetus three months before birth or three months after birth. In either case, that fetus's unique combination of genes is removed from the gene pool and will never have a chance to be passed on to future generations.

Second, those who favor abortion rights claim that the embryo/fetus is "part of the woman's body," so she should be able to decide what to do with it. But in terms of genetics, the embryo/fetus is not just "part of the woman's body" because, although half of its genes came from the mother, its set of genes is not identical to its mother's genes.

Dating and Marriage

While men can, in theory, go on fathering children into their fifties or sixties, women are at their most fertile—and most attractive to men—from the ages of about 15 to 25. However, in our society teenage marriage is frowned up and many women these days want to get a college education before getting married.

This means that after graduation from college a woman has only about 1000 days to "hook" a desirable (i.e., high status and thus well-adapted) man who will promise to commit to her for life. (In fact, the Japanese have an expression for a woman who has reached the age of 26 without getting married—a "Christmas cake"—referring to the fact that once December 25th has passed, such a cake has lost its attractiveness.)

Unfortunately, the influence of the women's liberation movement has made it more difficult to accomplish this. First, with women more "sexually liberated," there is less pressure on men to commit to a woman in order to get sex, and, with women more economically liberated, there is less financial pressure on them to get a husband. Secondly, with so much promiscuity, the rate of out-of-wedlock birth has skyrocketed and men are not eager to marry and become responsible for a woman who is responsible for another man's child (and his genes). Third, the women's liberation movement has denigrated marriage as a male-conceived institution whose purpose is to control, dis-empower and enslave women, and this makes out-of-wedlock birth seem less bad.

Summary

While Women's Liberation has undoubtedly expanded the opportunities for and increased the freedom of women in the short run, it has not been advantageous from our genes' point of view. It may seem overly obvious to say so, but the future belongs only to those who are alive to shape it and anything that artificially decreases the number of children women (and especially well-adapted women) have (especially to below the replacement level) is bad from an evolutionary point of view.

To liberals, conservative policies—such as those that are "pro-life," pro-traditional marriage, and anti-birth control—may seem regressive and anti-woman, but, as we have seen, they may in fact be much more in line with the laws of Evolution than liberal policies are.

Of course, the above discussion does not mean that women should be discriminated against or have their career options limited. (See Steven Pinker, pp. 350-354) However, just as it behooves us to think about how our present actions (such as in regard to so-called "climate change") will affect the future *physical* environment of our descendants, shouldn't we also think about how our present actions will affect the future *social* and *cultural* environment of our country and our descendants? And if we are willing to sacrifice economically to preserve the *physical* environment, shouldn't we be willing to sacrifice in other ways to preserve the *social* and *cultural* environment we live in?

"Universal Healthcare"[59]

According to a recent article on the British healthcare system:

> As of July 2024, 7.62 million patients were on the waiting list for care, with 6.39 million in need of specific medical treatment, according to the latest Referral to Treatment (RTT) data from the National Health Service (NHS), England's publicly funded health care system.
>
> The average wait time for treatment is 14 weeks, but more than three million patients have been waiting for over 18 weeks — and it's been more than a year for nearly 300,000 of them.
>
> Dr. Marc Siegel, senior medical analyst for Fox News and clinical professor of medicine at NYU Langone Medical Center, appeared on "Fox & Friends" to share his concerns about the situation.

---

[59] The following discussion on withholding healthcare from some people a) refers only to people whose health problems are caused by genetic factors, and b) is only theoretical and is offered only to make you see "universal healthcare" from a different point of view.

"This is a huge warning for us," he said.

As of July 2024, 7.62 million U.K. patients were on the waiting list for care, with 6.39 million in need of specific medical treatment.

"The National Health Service, which started in 1948 with the great idea to take care of everyone in England, is broken," he went on.

"We're talking about nearly eight million people there who are waiting for health care … many more than 18 weeks. How could you wait 18 weeks if you're having a heart problem or you have an infection?"

Although the problem is not as extreme in the U.S., Siegel warned that it can be a struggle to get timely care stateside.

"Even here … 26% of the people in the U.S. are waiting more than two months for their health care already," he told Fox News.

"Even people who are getting it from their employers are waiting."

But there are also long-term invisible (but inevitable) dangers of "universal healthcare."

The Evolutionary Cost of "Healthcare"

A problem arises when societies believe that "curing" health problems is better than allowing nature to decide who survives and passes on their genes to the next generation.

Let's take "Genetic Mutation Q" (GMQ) for example. For this example, we will say that one out of a thousand babies is born with GMQ.

A hundred years ago, any child born with GMQ would die before the age of ten. What a terribly sad thing for the child and a terribly painful thing for the parents! But *the deleterious mutation would die with him*! This is nature's way of eliminating deleterious genetic mutations from the gene pool.

Then, fifty years ago a vaccine was invented that allowed children born with GMQ to live near-normal lives. The vaccination costs $1000 a year, but it counteracts the effects of GMQ. This is great for the children born with GMQ and their parents!

But what are the negative consequences?

1) The "cost of living" goes up by a thousand dollars a year for these people, or about $75,000-$80,000 over their entire lifetime. In a socialized health care system, this means that the cost of health care goes up for *everyone*.

2) Not only the child survives but *the genes that cause GMQ* (which before the vaccine was invented would "die" with the child) also survive. And if those who have this mutation have children, there is a high probably that they will pass this malfunctioning gene on to their offspring.

Now, one in a thousand babies are still being born (due to a gene mutation) with GMQ, but, in addition, children of GMQ sufferers (who survived only because of the vaccine) are also being born. That means that, generation after generation, *the percentage of children born with GMQ will increase!*

Now, if you multiply this by all of the people who survive because of factors outside their bodies that they were not born with, you can see how this makes the cost of health care and survival rise significantly.

Another easily-to-understand example of this is eyeglasses (and contact lenses).

[G]enetics definitely play a role in whether a child will develop myopia. A child with two nearsighted parents can have a six times greater risk of myopia than a child with no nearsighted parent.[60]

A thousand years ago the percentage of people with poor eyesight was undoubtedly much lower than it is now because having poor eyesight would be a serious detriment to one's ability to survive. So fewer people with poor eyesight would survive, their "poor-eyesight genes" would die with them, and fewer people with "poor-eyesight genes" would be born.

Now, however, anyone with poor eyesight can easily get eyeglasses/contacts or get surgery to correct the problem. The result? Many more people with "poor-eyesight genes" survive and pass their "poor-eyesight genes" on to their children than in the past, resulting, once again, in more people having poor eyesight and creating more dependence on technological fixes (which may not always be there) outside their bodies and less on their genes, resulting in an increase in the cost of living. (See page 105 for a discussion of: "Vaccines: Good or Bad?")[iv]

Optometry researchers estimate that about half of the global population will need corrective lenses to offset myopia by 2050 if current rates continue – up from 23% in 2000 and less than 10% in some countries. The associated health care costs are huge.[61]

---

[60] "Is nearsighted genetic?"
[61] "Nearsightedness Rates Are Soaring. Here's Why"

An organism that can survive more "cheaply" (in terms of acquiring or making the energy and resources needed to survive) relative to how "expensive" (in terms of acquiring money, which represents a way of saving energy and resources to spend later) will have a higher chance of surviving. So, as least in the short term (in one generation), it would seem that people who are born with or can acquire/save more money will have a better chance of surviving and passing their genes on to their children.

However, if nature was in charge, if both a person who had a lot of resources and a person who had few resources had, say, eight children, more of the children of the person with more resources would survive than would the children of the person with fewer resources. That is just nature's way.

But a problem arises when people replace reliance on genes with external resources.

Let's take "Genetic Mutation X" (GMQ) for example. For this example, we will say that one out of a thousand babies is born with GMQ.

A hundred years ago, any child born with GMQ would die before the age of ten. What a terribly sad thing for the child and a terribly painful thing for the parents! But *the deleterious mutation would die with him*! This is nature's way of eliminating deleterious genetic mutations from the gene pool.

Then, fifty years ago a vaccine was invented that allowed children born with GMQ to live near-normal lives. The vaccination costs $1000 a year, but it counteracts the effects of GMQ. This is great for the children born with GMQ and their parents!

But what are the negative consequences?

1) The "cost of living" goes up by a thousand dollars a year for these people, or about $75,000-$80,000 over their entire lifetime. In a socialized health care system, this means that the cost of health care goes up for *everyone*.

2) Not only the child survives but *the genes that cause GMQ* (which before the vaccine was invented would "die" with the child) also survive. And if those who have this mutation have children, there is a high probably that they will pass this malfunctioning gene on to their offspring.

Now, one in a thousand babies are still being born (due to a gene mutation) with GMQ, but, in addition, children of GMQ sufferers (who survived only because of the vaccine) are also being born. That means that, generation after generation, *the percentage of children born with GMQ will increase!*

Now, if you multiply this by all of the people who survive because of factors outside their bodies that they were not born with, you can see how this makes the cost of health care and survival rise significantly.

Another easily-to-understand example of this is eyeglasses (and contact lenses). A thousand years ago the percentage of people with poor eyesight was undoubtedly much lower than it is now because having poor eyesight would be a serious detriment to one's ability to survive. So fewer people with poor eyesight would survive, their "poor-eyesight genes" would die with them, and fewer people with "poor-eyesight genes" would be born.

Now, however, anyone with poor eyesight can easily get eyeglasses/contacts or get surgery to correct the problem. The result? Many more people with "poor-eyesight genes" survive and pass their "poor-eyesight genes" on to their children than in the past, resulting, once again, in more people having poor eyesight and creating more depend-ence on technological fixes (which may not always be there) outside their bodies and less on their genes, resulting in an increase in the cost of living. (See page 105 for a discussion of: "Vaccines: Good or Bad?")

This leads to what I (humbly) call:

Showstack's Law

Anything outside the body that helps people survive is good in the short run, but people who would not have survived without it will survive and pass their "less-well-adapted" genes on to their descendants who will also not be able to survive without the outside help. The result is that, in the long run, it will increase the "cost of living" for everyone because people will become more dependent on (relatively expensive) things outside their bodies rather than the (relatively inexpensive) genes that they were born with.

Genetic Engineering

At first blush, the prospect of "genetically engineering" embryos to eliminate negative traits or transferring stem cells to the sick sounds exciting. No more disease! No more physically deformed babies! No more singers like Britney Spears!

However, there are at least four problems with the idea of having humans choose which genes babies will be born with.

1) By lowering the variation in the distribution of genes in a population, it makes it harder for the population to adapt to changes in the environment (and more susceptible to pandemics).

2) The genetic engineers may make a mistake. A trait which they see as maladaptive may have been adaptive in the past and may be adaptive in the future. Or the gene may never have been adaptive but may be so in the future.

A couple of examples:

If an embryo in the 1950s had contained the "nerd" gene, it might have been seen as maladaptive. After all, who needs more shy, sexually repressed guys with Attention Deficit Disorder who prefer staring at rectangular cathode ray tubes to staring at women's breasts? If the genetic engineers had been in charge, those genes might have been "deleted" and we would have missed the computer revolution of the 1970s and '80s! (See Miller and Kanazawa, pp. 26-27)

Another more fanciful example: Imagine there were aquatic "genetic engineers" at the time the first animals were developing the ability to breath air directly rather than extracting it from the water they were swimming in through their gills. The GE's might have thought, "Hmmm, this gene allows fish to survive on land for short periods of time—a useless trait. Out with it!" We might still be swimming around in the primordial sea!

A More Concrete Example!

Let's suppose that 99.9% of all mutations are bad. This means that of 1000 mutations, 999 are bad and one is good. And this means that out of 1000 people with mutations, 999 of them live worse lives because of their mutations but one will live a better life.

But which mutation is the one that is good? If a genetic engineer has his say, he will take the side of caution and "fix" all 1000 mutations. That will improve the lives of the 999 people who otherwise would have suffered with the mutation, but at the cost of perhaps losing the beneficial mutation forever. And, in the long term, the good done by the beneficial mutation as it is passed on to and spreads to more and more people might far outweigh the ill effects of the bad mutations on the 999 people who suffer with them.

3) Third, a gene which may "look bad" may in fact interact in some unsuspected way with another gene to allow that second gene to perform a useful function. One example is the

> [H]eterozygote advantage—where having two copies of a mutated gene can mean disaster but one copy is helpful.
>
> The most famous example of this is sickle cell anemia, which strikes people of African descent who have two defective copies of the hemoglobin B gene. As a result, they make red blood cells that are too curvy to carry oxygen to critical organs.
>
> People who have only one bad copy make useful red blood cells that are deformed just enough to

protect them from the malaria parasite, insulating them against the disease.[62]

4) A gene which may "look good" in one respect may have negative consequences in other areas. For example, one study found that while "women tend to prefer more toned men, and muscle-bound men tend to have more sexual partners than slender men, when other factors are controlled for," those muscles come at a price: a need for more energy input (i.e., food) plus "fewer infection fighting white blood cells and less of an important immune molecule called C reactive protein, which helps destroy pathogens."[59]

Pinker notes:

> Whether or not we *can* breed for certain traits, *should* we do it? It would require a government wise enough to know which traits to select, knowledgeable enough to know how to implement the breeding, and intrusive enough to encourage or coerce people's most intimate decisions. Few people in a democracy would grant their government that kind of power even if it did promise a better society in the future. The costs in freedom to individuals and in possible abuse by authorities are unacceptable.[63]

The point is that not only have our genomes evolved over millions years; the way that they change over time has also evolved. In other words, after millions, hundreds of

---

[62] Kaplan, K.
[63] Pinker, S.

millions, even billions of years of experimentation, Nature has reached a compromise solution to the problem of how evolution should work, a solution that maximizes the probability that genes will survive in the long run, and that solution includes random mutations and natural selection. This solution is not necessarily what's best for individual living things in the short term, but it is what is best as a way for the species as a whole to survive by adapting to their ever-changing environments in the long term.

Summary: Genetic Engineering/Stem Cell Research

As explained above, there is a real problem with trying to "genetically engineer people" because we can't be sure that what look like deleterious mutations now may not turn out to be advantageous in the future. Therefore, it is best left to random chance *even if this means that the vast majority of mutations people will be born with—and suffer from—will be deleterious.*

Socialism/Welfare

> [T]he welfare state is a very unnatural thing.
> Richard Dawkins, *The Selfish Gene* (p. 117)

The idea that anyone—but especially the poor—may suffer makes conservatives feel bad, but it makes liberals feel really really bad! Their solution is to "even the playing field" by taking money from those who have it and spending it on those who don't.

The problem is that this amounts to a "tax" on those who are so (born) well-adapted to their environment that they can live independent of outside help and a subsidy to those who are (born) so poorly adapted that they are dependent on outside help. And, as we have discussed in the section on Health Care, however bad it makes us feel that some people are doomed to suffer more in life than others in the short run, in the long run such taxes and subsidies do no good for they make it more difficult for the most well-adapted genes to be passed on to the next generation and easier for less-well-adapted genes to be passed on, and this just passes the problems on to the future.

There is a saying: "If you want more of something, subsidize it. If you want less of something, tax it." As Daniel Hannan wrote (in discussing the ideas of Iain Duncan Smith, a British politician who was a leader of the Conservative Party from 2001 to 2003) in a column entitled, "If you pay people to be poor, you'll never run out of poor people":

> Poverty is not simply an absence of money. Rather, it is bound up with a whole set of other circumstances: lack of qualifications, demoralization [sic], family break-up, substance abuse, father-lessness. It follows that you do not end poverty by giving money to the poor: a theory that British welfarism has amply demonstrated over 60 years. Only when you tackle poverty holistically will you facilitate meaningful improvement.[64]

---

[64] Hannan, D.

"The Theory of Evolution" necessarily includes "natural selection"—the idea that nature "selects" which individuals will win the competition to survive and reproduce and "survival of the fittest," which necessarily means that, without outside interference, the less fit (i.e., those who are less-well-adapted to the environment they find themselves in) will have a lower chance of surviving and passing their genes on to the next generation.

This may seem cruel. After all, what did the "less fit" do to deserve their fate? Aren't they as deserving of success and happiness as anyone else?

Unfortunately, Nature does not see it that way, nor does she care. She has the Bigger Picture in mind—the long-term survival of the group, which cannot occur unless the group adapts to its changing environment.

It is ironic that liberals, who claim to be supporters of the environment and of "The Theory of Evolution," do not understand this or, if they do, refuse to allow it to interfere with their desire to override Nature's plan.

Charity/Wealth

Charity

Charity (or "forced charity," i.e., welfare) used to be make sense when mankind lived in small clans who were all genetically related to each other. However, it does not make (biological) sense to give money to people with whom we are not related because it does not serve to our own genes.

But don't the better adapted have a moral duty to help the less well adapted?

The above, of course, brings up all sorts of moral and social issues. Is it moral for people to have more children than they can raise? If they cannot raise them, who will? How can we stand by and let "mal-adapted" people die? Who knows what birth rate is best? What about the effect on the environment of an increasing population? Should a couple who decide to have one or two children and devote all their resources to raising them be taxed in order to subsidize a family which has eight or ten children? And which is better: to have five children, only three of whom survive or to have three children, only two of whom survive?

Summary: Socialism/Welfare/Charity

As discussed above, socialism taxes those who do better in society and subsidizes those who are less successful. As we have seen, while this may seem fair and feel good in the short run, in the long run over many generations it helps genes survive which make certain people *less well adapted, less independent and therefore more and more in need of outside assistance in order to survive*. In other words, welfare does not cure poverty, it perpetuates it.

Conservatives prefer charity to welfare. One of the advantages of charity over government redistribution of wealth is that, with charity, the giver feels a sense of fulfillment and the receiver (at least in theory) feels grateful. With government welfare, on the other hand, the "giver" (who is forced to pay taxes) may feel resentful while the recipient feels "entitled."

Another difference between the two is that with welfare the decision on who is to receive aid is not based on any moral or behavioral conditions. With charity, on the other hand, the giver can give money to those whom he wants to help and encourage, so the giver can set some terms the recipient must meet if he or she is to receive the charity.

Because they are lenient, lack the desire to enforce strict standards of behavior and enforce high moral values, liberals see themselves as kinder and more generous and tolerant than conservatives, who they see as mean, heartless, self-centered and interested in controlling other people's lives. However, surveys (in the U.S.) have shown that, on a percentage basis, liberals are markedly less charitable than conservatives, who give a higher proportion of their income to charity than liberals do and also volunteer more hours to charity and community service.[v]

Using Charity to Shape Society

It should be obvious that charity, which allows a person to personally choose which people or organization is to receive his money, better allows a person to help those with similar values to survive and pass their genes on to the next generation than government-sponsored welfare programs do. This, in turn, helps shape society in a way that is more to the giver's liking, which means that it will likely be a better "environment" for those he passes his genes on to.

Of course, if we took this to an extreme, it would mean allowing people who are "mal-adapted" to living in our society to simply die. While Nature would not object to this because it enables the society as a whole to be better adapted to the environment, it is some-thing that few people would want or allow. Even Darwin would not countenance it:

> The aid which we feel impelled to give to the helpless is mainly an incidental result of the instinct of sympathy, which was originally acquired as part of the social instincts, but subsequently rendered, in the manner previously indicated, more tender and more widely diffused. Nor could we check our sympathy, even at the urging of hard reason, without deterioration in the noblest part of our nature. The surgeon may harden himself whilst performing an operation, for he knows that he is acting for the good of his patient; but if we were intentionally to neglect the weak and helpless, it could only be for a contingent benefit, with an overwhelming present evil.[65]

In addition, just because an individual is not able to survive on his own does not mean that his children will be equally maladapted.

However, if the government spent less money on welfare, it would take less from those who earn it and more money would be available to private citizens to give to charity (or have children).

---

[65] Darwin, C.

Is Wealth Correlated with Adaptiveness?

We tend to think that a person's "adaptiveness" (ability to survive and pass one's genes on to one's descendants) is correlated with wealth, but today that is not necessarily the case.

But that is not the case today, for two reasons. First, because of "progressive" social policies and beliefs, more people are choosing to have fewer children than they could afford to raise. Second, again, because of "progressive" social policies and beliefs, people who are not able to provide for children without help are going ahead and having more children than they can afford because they are receiving various forms of aid and welfare from the government. And, as is the case with medical care (that people who cannot survive without outside help are helped to survive by outside care, the result of which is that they are also passing on their "can't-survive-without-outside-medical-help" genes to the next generation), people who cannot survive financially *for genetic reasons* without outside help are being enabled to survive with help from the government, the result of which is that they are also passing on their "can't-survive-without-outside-financial-help" genes to the next generation.

Let's look at it another way. Let's say that one person has four times as much money as another person. And let's say that both of them have ten children. If nature was allowed to take its course, eight of the children of the rich person would survive and have children but only two of the children of the poor person would do so. And the result? Each of the two surviving children of the poor person *would inherit the same amount of money* as the eight

93

surviving children of the rich person! In other words, nature has a way of "evening the playing field" that results in narrowing the wealth gaps in the next generation, but people today do not want to play by her rules.

The Biological "Cost of Living"

In order for an organism to survive, it must take in nutrients, convert those nutrients into energy and use that energy to accomplish what it needs to do to survive. As discussed above, organisms with a lower "cost of living" will have a selective advantage in the competition for survival over those with a higher "cost of living," and *there is a high correlation between "cost of living" and dependence on outside resources: organisms which have the least need for outside resources will have the best chance of surviving.* (Remember this later when we discuss liberal social programs.)

Unfortunately, over generations, medical care *raises* people's "cost of living" because, as people become more and more dependent on drugs, treatments, etc., to do what their maladaptive genes cannot do, their "cost of living" goes up.

So what would happen if we stopped all medical entirely for genetic diseases? Obviously, a lot of people (maybe most people—including me!) would die! But the survivors would be the ones who have genomes that are so well adapted to the environment they live in that they do not need medical care and they would pass their healthy genes on to their descendants. (Of course, because of mutations to their genes and changes to their environment, some of their descendants will not survive to reproduce, but that is nature's way of maintaining a well-adapted gene pool.)

To help us think about this, let's look to the past—to the plagues that devastated Europe in the Middle Ages.

Let's suppose that when the first plague was brought to Europe, Brainius Maximus, a brilliant medieval alchemist/doctor/scientist, had come up with a vaccine or cure for it. Everyone in Europe would have been vaccinated against it. And what would be the result? For one thing, fewer people would have died from the plague and more would have survived. However, it would have entailed a tremendous cost *in the long run* because *everyone* since then would have had to continue to be vaccinated against the plague. Secondly, we might still be in the "Dark Ages"! That's because (as many historians believe) the plagues, by decimating the work force and thus raising the cost of labor, changed the economic relationships in society to such an extent that it led to the rise of the merchants, the middle class and the Renaissance.

Instead, what happened? Many many people who were susceptible to the plague suffered horrible deaths, but those who had greater natural resistance to it survived, and almost all of us alive today who are of European descent still have some natural resistance to the plague because we are descended from those who survived it.

In addition, there is a continual "arms race" between us and the pathogens that wish to attack us. As one after another of the "bad" bacteria have developed resistance to our antibacterial drugs, we have had to try to develop new ones that will (temporarily) kill them. But, again, each time we "defeat" a microbe, we are allowing people to live who then may pass on genes to descendants who will need to be protected against those microbes.

Not curing people with genetic diseases (or at least keeping them alive) when we have the means to do so? What a horrible idea! Unfortunately, that "horrible idea" is the way Nature operates.

Let's do a thought experiment:

What if, in two identical societies, only one continues to treat its citizens for genetically-caused illnesses and the other doesn't. How would the two societies compare after thousands of years?

The citizens in the two societies might be equally "healthy" but it would cost a lot more money to keep the citizens in the "treatment intervention" society healthy than in the "don't treat" society because the citizens in the latter would all be descendants of people who had survived without medical treatment and thus would have more natural immunity to diseases. That would mean that the general standard of living of the people in the "treatment intervention" society would be lower than the standard of living of the people in the "don't treat" society because people in the "treatment intervention" society would have to spend increasing amounts of money on treating people for health problems that have been eliminated (through premature death) in the "don't treat" society.

If you ask yourself, How can we survive without medical care?, then I would suggest you look at all of the millions of wild species on earth which are surviving quite well without it.

Is there any possibility that society will adopt a policy to end all medical care for genetically caused diseases? Not a chance!

For one thing, how do we determine what is and what is not a genetically caused disease or injury? If someone gets drunk and gets injured in a car crash, is that a genetically related injury because he was genetically susceptible to becoming an alcoholic? If someone gets mugged while taking a midnight jog in Central Park, is that genetically caused because he is a night owl who suffers from insomnia and is genetically programmed to like to exercise and to take chances, like jogging at midnight in Central Park?

Of course, the main reason we will never adopt such a policy is that we, in our short-term outlook and immediate concern for the health and survival of ourselves and others (especially those who share our genes), will never condone it or allow it.

The problem is that eventually an outside force (Global warming? A new ice age? A nuclear war? A massive volcanic explosion? The earth being hit by a wayward asteroid? Or what about a massive burst of electro-magnetic radiation from the heavens that would wipe out our entire power grid for weeks, months, or even years?) is going to make "the cost of living" (that is, the cost of staying alive) go up to the point where it will not be possible to keep those dependent on expensive medical interventions alive.

If (or perhaps we should say, "when") such a worldwide natural calamity occurs, who will be more likely to survive—people in highly "advanced" countries like the United States who depend on highly technically advanced medical care (treatments and medication) or people who live in "backward" societies without advanced medical care? The answer is the latter group of people will have a better chance of surviving because, although they have a "lower standard of living," they will not be so seriously affected since they are not dependent on advanced medical care, which will no longer be available or affordable.

This brings up one other issue: those people who, for religious reasons, refuse to give medical care to their children. In "advanced" societies such as the United States, those people are usually charged with a crime and their children are given medical care to keep them alive regardless of the wishes of their parents. However, as we have discussed above, although refusing medical care to children may seem morally outrageous—and is certainly socially unacceptable—to us "modern thinking" people, there is a *genetic* argument to made that allowing children who cannot live without medical care to die makes sense, at least on the long-term genetic level.

Why "Universal Health Care" Makes a Society *Less* Healthy

Liberals favor "universal health care" because they believe that health care is a basic human right, so it would no doubt surprise them to learn that, in the long run, *medical care (at least for genetically caused diseases) doesn't do any good!* That's right: medical care has not made it any more likely that the human race will survive. In fact, it may have done *the opposite*. This is because saving the lives of those who would have otherwise died from a genetically caused disease raises the cost of keeping future generations alive by weakening the health of society at large and also takes money away from what it could be devoted to, which is raising the next generation of healthy genetically well-adapted children.

As Arnhart observes:

> [W]hile among savages those individuals who are physically or mentally weak are eliminated, and only the most vigorous survive, civilized societies impede this process of elimination, because the sick and impaired are cared for. "Thus the weak members of civilized societies propagate their kind," Darwin observed, and "no one who has attended to the breeding of domestic animals will doubt that this must be highly injurious to the race of man."[66]

In other words, Nature's way is to invest in protecting and nurturing the next generation rather than trying to treat or "cure" people who may die because of their condition

This is not to suggest that we should treat humans like farm animals. It is only to once again to point out the irony that it is Liberals who are the primary defenders of Darwin and his views and yet favor social policies that show no understanding of how evolution works.

Some Concrete Examples

Let's divide the situation into three parts. (And, once again, I am not saying that the following is "good" or "the

---

[66] Arnhart, L.

way it should be." I am only pointing out that it is the way it is. Nature is indifferent to our feelings.)

A) If we save the life of a child who otherwise would have died from a genetic disease, that allows that child to grow up and possibly pass his flawed genes on to the next generation where health care will again have to be relied upon to keep his descendants alive.

Think of all of the childhood diseases that are now "cured" either by medical care or prevented entirely by vaccinations. That's great for the children who get to live out their lives, but, as time passes and their maladapted genes are passed on, the general health of society at large deteriorates as more people become more and more dependent on outside cures and procedures to survive.

B) Next, let's consider the case of medical care for those aged 15 to 70 years old. Medical care (treatments and cures) will help those with genetic diseases to survive who otherwise wouldn't survive, which means they will be able to have children (and grandchildren), whom they can help raise. The problem is that they may pass their maladapted genes on to their descendants, which again will mean that, over time, more and more people will be less and less healthy (that is, less able to survive without outside intervention), which makes it more and more expensive to give "universal health care" to society at large.

C) Now let's talk about the sensitive subject of healthcare for those over the age of 70, the time on earth ("three score and ten years") that was allotted to us even as far back as the time the Bible was written. (By the way, it is not true that as people have become "healthier" over the centuries that people are living longer. In fact, the "maximum life span" has not changed much for millennia. The reasons that the "average life span" has risen is that now many fewer children die in childhood and many fewer women die in childbirth, mostly due to better sanitation.)

Humans have long dreamed of creating an elixir that would extend their lives, even make them immortal. However, attempting to artificially lengthen life span conflicts with nature's goal of keeping the DNA "alive."

Why do humans generally "shuffle off this mortal coil" after seven decades or so? It's because Nature has determined, after hundreds of thousands of years of experimentation on hundreds and hundreds of generations of humans, that that is the optimal life span for humans in order to maximize the survival potential of *their genes*.

Once people hit around the age of seventy, however, they start to take more resources from their families—and society—than they contribute.

From the point of view of our genes, anyone who is not contributing to the goal of making future copies of themselves is a deficit (to the genes). Sadly, there comes a time in every person's life (and that time is usually around the age of 70) when a person starts to use up more resources that he or she can contribute to society. For this reason, I take every article about "extending the life span" with a grain of salt (which I shouldn't do because of my problem with high blood pressure!) because I know it is doomed to fail.

But what if someone came up with an elixir that could extend people's lifespans to 100, 150, even two hundred years? It would be a disaster for mankind! Remember: Nature knows what she is doing.

Even liberals now have to face the fact that the promises we have made to ourselves to care for people in their old age through Medicare and Social Security will be hard to keep. And remember that every dollar spent on keeping an old person alive is one less dollar spent on care for the younger generation.

In addition, as time goes on, more and more money is spent on curing or managing diseases that are more and more rare. (And, with improvements in technology, more and more rare diseases can be detected and treated.) And while it may cost $20 a person to cure or prevent a common disease, it may cost $200 a person to cure or prevent a slightly less common disease, $2000 a person for a less common disease, and so on. That is another reason the cost of healthcare is going up. (A recent article in the *New York Times* with the headline, "The $6 Million Drug Claim," notes that, "New treatments for rare diseases are changing the lives of patients, but the price can reach millions of dollars for a single person."[vi]

Conclusion: Why Socialized Medicine is (Genetically) Unhealthy for Society

People should always retain the ability (within limits) to do what they feel is necessary to pass their genes on to the next generation. It makes sense (genetically speaking) for individuals to try to stay alive and to save the lives of those who are genetically closely related to them because it increases the chance that their shared genes will be passed on to later generations.

However, socialized medicine interferes with this ability because it forces people to contribute to the survival of the genes of people to whom they are not closely related, genetically speaking. It does not make sense for the government to tax genetically well-adapted (more successful) people in order to provide health care to others who are not genetically well-adapted and are not (genetically) closely related to those being taxed. All it does is make it harder for well-adapted people to pass their own genes on and make it easier for the less-well-adapted to pass their genes on to future generations. (See the section above on "The Long-term Results of Women's Liberation.") It is, in other words, the biological equivalent of "deficit spending"—expecting future generations to pay higher health care costs in the future so that we can be happy, healthy and comfortable now.

The other problem with taxing the better adapted in order to subsidize the less-well adapted is (as explained above in the section on "How Evolution, Natural Selection and Adaptation Work") that it interferes with the ability of a population to adapt to an ever-changing environment, and a population that does not adapt will not survive.

But (you may ask), doesn't the government have *a moral duty* to provide health insurance to everyone?

The surprising answer is that even *health insurance* may not make a difference. In an article entitled "Myth Diagnosis," Megan McArdle notes "the results of what may be the largest and most comprehensive analysis yet done of the effect of [health] insurance on mortality....In test after test, [the researcher, Richard Kronick of the University of California at San Diego] found no significantly elevated risk of death among the uninsured."[vii]

And here is an excerpt from an article titled, "Obamacare Was Supposed to Lower the Death Rate—It Didn't":

Prior to the passage of the Affordable Care Act, Democrats and their policy-group echo chamber asserted that tens of thousands of Americans were dying every year because they didn't have health insurance. Of course, now they're saying "millions and millions will die" because of climate change.

For the uninsured dying, Democrats pointed to an Institute of Medicine (IOM) study in 2002, based on a 1993 study, asserting that some 18,000 people died annually due to a lack of health insurance.

Then in September 2009, as the battle over Obamacare legislation was raging, the American Journal of Public Health released a study, conducted at Harvard Medical School and Cambridge Health Alliance, that—surprise!—significantly increased the IOM estimate.

The Journal claimed that 45,000 Americans died annually due to a lack of health insurance.

Of course, the ACA became law, and the Obamacare exchanges became operational in 2014.

As a result, the number of uninsured dropped significantly—from 17.8 percent in 2010 to 10 percent in 2016, with a slight uptick to 10.2 percent in 2018, according to the Kaiser Family Foundation. Most of the decline was due to Medicaid expansion.

Since the health insurance expansion we have seen the death rate decline every … oh, wait a minute!

No, in fact the U.S. death rate has grown since … 2009.[67]

---

[67] Matthews, M.

As the Trading Economics graph shows, the crude death rate was 7.9 persons per 1,000 people in 2009 and 8.4 per 1,000 in 2016. If having health insurance was such an important factor in preventing deaths, shouldn't the death rate have declined, or least remained the same?

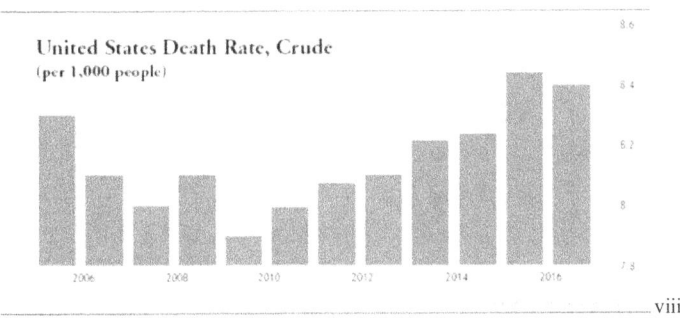

United States Death Rate, Crude
(per 1,000 people)

viii

In summary, those who truly believe in Evolution should oppose "universal health care" because, while everyone should have the right to do whatever he or she feels is necessary to maximize the chances that his own genes will survive, it is evolutionarily deleterious for the government to force those who have better adapted genes to use their resources to help those with less-well-adapted genes survive and reproduce.

In short, in the long run, socialized medicine makes a society less healthy.

Vaccines: Good or Bad?

This leads to the touchy subject of *vaccines*.

Recently, due to the high number of measles outbreaks occurring all over the world, there has been much discussion/disagreement/debate about vaccines.

I was vaccinated and I made sure to have all of my children vaccinated. But what is the *long-term benefit/effect* of vaccines to society as a whole?

As explained above, allowing people to survive who would not have survived without the help of a vaccine means that their "can't-survive-disease-X" genes also survive. If the vaccine had not been invented, those people would have died and would have taken their "can't-survive-disease-X" genes with them, and those who had a natural immunity to the disease would survive. This is nature's way of fighting the disease. But because of the vaccine, the population/species will become more and more dependent on it, and what happens when, someday, it is no longer available?

Let's take a more quotidian example: air conditioning.

Before air conditioning was invented, the number of places people could live and work was limited by the climate, and people died because of excessively hot weather. But they took their "can't-survive-excessively-hot-weather" genes with them, and the population as a whole remained adapted to the place and biome in which they lived.

However, once AC was invented, it: a) allowed people to survive who otherwise would not have been able to survive, and b) allowed them to live and work in environments which they previously would not have been able to live. (Good outcomes.) *However*, it once again: a) made the "cost of living" more expensive because AC became *necessary* for some people to survive, and b) made people more dependent on things outside their bodies rather than the genes they were born with.

To sum up: Helping people survive by means outside their bodies (such as eyeglasses, vaccines and AC): a) increases the "cost of living"; b) makes people more dependent on things outside their bodies; and c) will make it more difficult for them to survive if the "thing outside their body" becomes more expensive or ceases to be available.

An organism which must "spend" more than it can "make" will not survive without outside help, and (according to Nature's logic) we should not help them to survive because, if we do, we will not only be saving the life of the organism but also saving the "lives" of the organism's less well adapted genes and allowing the organism to pass those genes onto the next generation.

But what do we do about the "less-well adapted" members of society.

That's a tough question to answer. As explained above, forcefully taking money from the better adapted members of society to help the less well adapted ones soothes our consciences in the short run has negative consequences in the long run.

Perhaps people who worry the most about "disadvantaged" people in "lower socioeconomic" classes should be encouraged to give them more money in the form of charity.

Summary: Implications and Ramifications

As we have seen, in area after area the beliefs, policies and programs of conservatives, even religious conservatives, are more in harmony with the way Evolution works than the beliefs, policies and programs of liberals are.

(Incidentally, another irony is the different ways liberals and conservatives are reacting to the prospect of "climate change." Liberals, who claim to want change in society, act terrified of any change to our climate while conservatives react to the prospect with more equanimity, believing that human beings will be able to adapt to such a change [if it ever happens].)

Now some liberal critics may argue that all these arguments are moot because humans are no longer evolving. However, a recent study found that, "The pace of human evolution has been increasing at a stunning rate since our ancestors began spreading through Europe, Asia and Africa 40,000 years ago, quickening to 100 times historical levels after agriculture became widespread...."[68]

It is often assumed that modern humans are no longer evolving. But there is now considerable agreement among scientists that evolution is still affecting our species—and this process is taking place "more rapidly" than ever before....

While cultural and technological innovations now appear to be the main drivers of adaptation for modern humans, this has not replaced biological adaptation, according to scientists.

"I don't think [the question of whether humans are still evolving] is fully appreciated by the general public," Michael Granatosky, an evolutionary

───────────────

[68] "Study finds humans still evolving, and quickly."

biomechanist at the New York Institute of Technology, told Newsweek....

"[E]volution simply means a change in a population's gene pool over successive generations. With this broader definition, I do not believe there is considerable debate among evolutionary biologists that humans are still evolving," he said.[69]

Nature believes in random variation, not equality.

"Woke" people today are demanding more than equality of opportunity; they demand "equality of outcomes," or "equity." ("Equality" means each individual or group of people is given the same resources or opportunities. "Equity, on the other hand, states that people are born in different circumstances and resources and opportunities need to be allocated in a way that results in equal outcomes.)

Unfortunately, this lies in direct conflict to what we know about the reality of human biology how evolution works.

There are two bases to "The Theory of Evolution."

First, not only is every person alive today genetically different from any other person alive today, but each individual is *unique*, genetically different from any other person who has ever lived or who will ever live!

Second, as we have seen, adaptation requires that some (better-adapted) individuals have a better chance of surviving and passing there genes on to the next generation

---

[69] "Are Humans Still Evolving?"

than individuals who find themselves in a changed environment which they may not be well-adapted to.

Of course, this does not justify discrimination towards any group of people by another group of people who believe they are less deserving of equal rights, opportunity and survival.

Remember*: Nature Always Wins in the End*

Once again, I want to emphasize that the purpose of this book is *descriptive*, not *prescriptive*: to make people think about the long-term effectiveness and results of various social policies. Isn't understanding these natural forces better than ignoring them.

Summary of what we have learned

1) Nature does not care about you or any "living thing." All she cares about is helping the most valuable thing in the universe — the genetic code to continue to exist.

2) An organisms environment is always changing. The environment you lived in yesterday is different from the environment you live in today, and, similarly, the environment you will live in tomorrow will be different than the environment you live in today. So the only way for the genetic code for a group of related organisms to survive is for the group to adapt to the everchanging environment.

3) In order to achieve number two, it is necessary for a group of organisms to *produce more offspring than can possibly survive*. This is necessary to Nature can "choose" the ones that are better adapted to the new environment to survive and bring forth the next generation. This means that it is Nature's plan that *not all organisms survive*.

4) Any attempt to challenge the above is guaranteed to meet with failure, sooner or later. In fact, trying to put short-term "anti-natural" measures in place to fight or work around nature's plan will (as explained earlier) make the likelihood of survival *less* likely.

Now you may say: This is nonsense! I know what I think and perceive and feel. However, the truth is that the ways you think and feel and perceive the world have evolved just like the rest of you. Those people who think and feel and perceive the world in a way that increases the likelihood that they will pass on their genes to the next generation will be more likely to do so than people who think and feel and perceive the world in ways that lower their chances of passing on their genes.

Nature does not care about "truth"! The *only thing* it cares about is keeping the genetic code alive. This is why people who dismiss people's religious beliefs need to read the next section.

People are free to choose and support whatever social policies they think best but, as has been shown above, no matter what people do, Nature *always* wins in the end.

Letting Nature be in charge may be a horrifying thought, but just because something is horrifying does not make it untrue or mean that denying its verity leads to a better strategy for survival.

Just remember: It is *always* a bad idea to doubt nature's wisdom.

Part Three
Other Things About Our Lives
That an Understanding of Evolution
Can Help Us Understand

"Is Evolution antithetical to morality."

How Morality, Traditional Religious Beliefs and Conservative Standards *Evolved Naturally*

[G]oodness can evolve, at least when the appropriate conditions are met.

[T]here are two very different pathways to evolutionary success. One involves exploiting your neighbor and the other working with your neighbor to achieve jointly beneficial outcomes. The second pathway provides room in evolutionary theory for what we call goodness.

A gene is called selfish whenever it survives and reproduces better than other genes, all things considered, but that is just newspeak for "anything that evolves." For example, suppose that the traits associated with good and evil... have a genetic basis. If the good traits evolve... the Richard [Dawkins] would call them selfish because they replaced the evil traits. [70]

Much of the philosophical debate surrounding evil is a question of human nature...Are humans, as a species, inherently capable of evil? Are we predisposed to it? Is it unavoidable? The answers to these questions might be informed by the science on how life (and our human brains) evolved to maximize survival and reproduction, especially among species that live within social groups....[T]he human species evolved the capacity for doing things we now regard as evil because, under certain conditions, those

---

[70] Wilson, D.S.

evil things increased the likelihood of survival and reproduction for an individual and its genetic offspring.[71]

Altruism

If our genes are "selfish," why are people nice to each other?

One reason (as explained above) is that one way to pass on copies of our genes is to help others who share our genes to survive.

And the eminent ethologist Richard Dawkins has another reason:

"[G]ne selfishness will usually give rise to selfishness in individual behaviour. However, as we shall see, there are special circumstances in which a gene can achieve its own selfish goals best by fostering a limited form of altruism at the level of individual animals."[72]

So where did morality come from?

If evolution is about survival of the fittest, how did humans ever become moral creatures? If evolution is each individual maximizing their own fitness, how did humans come to feel that they really ought to help others and be fair to them?

There have traditionally been two answers to such questions. First, it makes sense for individuals to help their kin, with whom they share genes, a process known as

---

[71] "The Evolution of Evil"
[72] "The Selfish Gene"

inclusive fitness. Second, situations of reciprocity can arise in which I scratch your back and you scratch mine and we both benefit in the long run.

But morality is not just about being nice to kin in the manner that bees and ants cooperate in acts of inclusive fitness. And reciprocity is a risky proposition because at any point one individual can benefit and go home, leaving the other in the lurch. Moreover, neither of these traditional explanations gets at what is arguably the essence of human morality—the sense of obligation that human beings feel toward one another.

Recently a new approach to looking at the problem of morality has come to the fore. The key insight is a recognition that individuals who live in a social group in which everyone depends on everyone else for their survival and well-being operate with a specific kind of logic. In this logic of interdependence, as we may call it, if I depend on you, then it is in my interest to help ensure your well-being. More generally, if we all depend on one another, then we must all take care of one another.[73]

So how did morality *evolve*?
Answer: In the same way that anything evolves!
Explanation:
Homo sapiens, the first modern humans, evolved from their early hominid predecessors between 200,000 and 300,000 years ago. By about 100,000 years ago, there were about one million humans roaming around the earth, and they started to give up their nomadic lifestyle and settle

---

[73] Tomasello, T.

down in permanent agricultural societies about eight-to-twelve thousand years ago. These first societies were composed of "clans," people who were related to each other. If there were (on average) about 100 people in each of these communities, then there were about 1000 of these early groups of people living together.

Then they had to figure what rules they should live by in order to live together and survive.

Now, for the sake of discussion, let's say each society had to agree on the answers to **ten basic questions**, such as:

—Religion: Shall we worship one god, many gods, or no gods?

—How should we decide who makes decisions for our group, and how much power should they have?

—How should we defend ourselves against enemies and when should we attack first?

—Which things should be outlawed in our society and which should be allowed?

—How should we punish people who break the laws of our societies?

—Should each family be allowed to keep the "profits" of their labor or should they be forced to share them?

—What forms of sexual behavior and relationships should be allowed and what forms should be banned?

—What sorts of ceremonies, celebrations, holidays, etc., should we have?

—Who should be in charge of raising children?

—How should we deal with disputes between people?

And so on….

A little math shows that there are about a thousand possible groups of answers to these questions, so it was possible that each of the thousand groups of people had a unique set of "morals" they chose to live by.

And then what happened?

Evolution!

The chance that an individual tribe survived depended on how close their arbitrary rules aligned with nature's rules — tribes which, for example, favor human sacrifice would have a lower likelihood of surviving than those which didn't.

The societies whose rules better allowed them to survive in their environment would have more offspring and would survive longer, and the societies who chose rules that were less well adapted to helping them survive in their environment would do worse.

Of course, over time, some of the more successful societies would split into two or more societies, some societies would conquer (or be conquered by) other societies, some would suffer from natural disasters, etc., and all of them would have to learn to survive in new conditions as their particular environment changed over time.

You can see where this is going:

Overtime, certain sets of moral rules would survive and less-well-adapted sets of moral rules would die out.

And that's how morality evolved!

A Defense of Religion and Traditional Mores

Nowadays, religion and traditional ways of doing things are considered not just passé but politically incorrect.

How can anybody defend those white European misogynistic patriarchal racist sexist ageist ableist (I'll let you fill in the rest of the ever-lengthening list of terms)?

Well, I am, now.

Maybe Religiosity isn't So Stupid After All

People on the left of the political spectrum make of fun of people who believe in "silly myths" of religions, but this just shows how little people on the left understand Evolution (and religion).

Religions evolve just like any other "living thing."[74] Those groups who adopted religious beliefs that were more in line with the rules of evolution would be more likely to survive and pass on their religious beliefs to their descendants than those who adopted religious beliefs that were not close to evolution's rules. (Remember: Nature doesn't judge people's moral or religious preferences. All she cares about is the survival of the genetic code.)

But what are these rules? Nobody knows!

So the only for societies — and religions — to survive is for people to protect the environment, protect their culture, and have lots of children.

And Maybe Conservatism isn't So Stupid Either!

Why do conservatives (and conservative religions) condemn anal sex, birth control, abortion, homosexuality, oral sex, "fornication," pornography, masturbation, etc., as sinful? The simple answer is that the Holy Books say it is so. But what would "Nature's Book" say about it?

---

[74] See the discussion on the "ten basic questions," on page 116.

The fact is that all of these behaviors interfere with nature's "Prime Directive," which (in Biblical language) is to "be fruitful and multiply." As discussed above, Nature wants us to do only one thing: pass copies of our genes on to our sons and daughters and help them (or others who share our genes) survive so they, in turn, will pass their genes on to their children. Even to Nature, anything which prevents or interferes with this is biologically "sinful" in the sense that it prevents us from passing on our genes (or makes it less likely that we will do so). So, what the "Holy Books" tell us in this regard is not so different from what "Nature's Book" tells us!

Now do you see why traditional religions and conservative social ideas are more natural than the modern variations on nature's plan?

Remember that all nature cares about is keeping the genetic code alive, and anything that interferes with that goal is "anti-natural." And what is nature's Rule Number One to make that happen?

All groups of organisms must have more offspring than can possibly survive.

And what are some things that interfere with Rule Number One?

Abortion, birth control, homosexuality, masturbation (and pornography), and so on.

But don't women (and men) have a right to decide what to do with their own bodies?

To which I answer: If abortion and birth control were in nature's plan, she would have given women the natural ability to avoid getting pregnant and to abort a pregnancy if it is not wanted.

As for the latter (homosexuality, masturbation and pornography), all of these misuse the natural (biological) abilities people are given by providing other ways to experience sexual pleasure other than sexual intercourse between men and women.

So traditional religious and conservative rules *evolved naturally through natural selection*!

Some people will argue, "That stuff made sense in the past, but now, with the modern technology and scientific knowledge we have, we don't have to follow those old-fashioned ways. We now have modern technology."

BUZZ — WRONG ANSWER! (and it also wasn't phrased in the form of a question).

There are two reason this answer is wrong.

First, no one can predict what kinds of natural (or man-made) disasters will strike us in the future.

Example one: What if a meteor the size of the meteor that led to the extinction of the dinosaurs struck earth?

Example two: What if a super volcano erupted and spewed so much stuff into the atmosphere that no sunlight reached the earth's surface for years?[75]

So how many children do we need to have in order to survive such a disaster?

Once again, the answer is, nobody knows!

Nature spent more than *three billion years* testing and perfecting (and adapting) the rules of life before the first

---

[75] This actually happened: "Modern scholarship has determined that in early AD 536 (or possibly late 535), an eruption ejected massive amounts of sulfate aerosols into the atmosphere, which reduced the solar radiation reaching the Earth's surface and cooled the atmosphere for several years." (Wikipedia)

hairy *homo sapiens* appeared on the scene. Don't you think that she has learned, through trial and error) the best way for life to survive?

In short, morality pushes up to do things that are beneficial to the survival of our genes away from things that are detrimental to that.

War: What is it good for?

It has become comforting among people on the wrong side of the political spectrum to muse things such as, "Give peace a chance," "End war," and all of those other banalities you will see on banners, T-shirts and placards that leftists smugly display.

Unfortunately, war will never go away because (if you win) it is an efficient way to obtain more land and resources and (for men) to capture more women (and girls) who can be used to pass on your genes to the next generation.

(It is estimated that Genghis Khan may be the ancestor of as many as 16 million descendants, and about eight percent of Asians today are descended from him!)

In fact (and I hope this doesn't disgust you), the invention of the atomic bomb has changed war (and societies) in two ways. The good way is that we have not had an all-out war since then because all of the countries with nukes are afraid to use them lest they be nuked themselves.

But another effect is the way it has affected societies. Human societies evolved in the environment of wars (or the constant threat of war) for millennia, and this meant that young men would be asked to fight. This meant that men evolved to have aggressive tendencies that needed to be channeled in a positive way when not at war.

Now we have to figure out a way for young men to use the testosterone that they evolved to have in a way that is not deleterious to the society they lived in. But, now, many (single) young men have few positive ways to take use their testosterone.

Secondly, war was a way save less-advantaged young men from a purposeful existence and bleak future and instead "make a man" (or even a hero) of them. (Of course, a bad effect of this is that many young men would be killed or maimed for life, but it also, conversely, had some positive effects.)

As Thomas Sowell said, "There are no solutions in life. There are only tradeoffs."

Speaking of Young Men and Excess Testosterone: Terrorism

Now let's consider something that is a little bit off the subject but serves to show how powerful genetics (and social rules) are in affecting behavior.

One subject that is on everyone's minds these days is terrorism, and would you believe that genetics and social rules help create terrorists?

Here's why.

According to one study, "a healthy marriage may decrease a man's propensity to commit crimes."[76] And, "The simple fact is that married people are less likely to commit crime than single people. And, offenders who

---

[76] Kirchner, L.

marry are less likely to reoffend, even if they have had long criminal careers." And, finally, unmarried men have "a propensity for forming gangs, joining violent political groups, and being three times as likely to commit murder than a married man of the same age."[77]

But what does that have to do with terrorism?

Because of the environment they lived in, ancient Middle Estern societies evolved as "clans" (groups of close-knit and interrelated families) or "tribes" (social groups made up of many families, clans, or generations that share the same language, customs, and beliefs, typically having a recognized leader. surrounded by other clans).

Society was patriarchal, with inheritance through the male lines. Tribes provided a means of protection for its members; death to one clan member meant brutal retaliation. *Non-members of the tribe were viewed as outsiders or enemies.*[78] [Italics added]

This diagram can help visualize their situation:

```
              clan   clan

      clan   clan   clan   clan

          clan   clan   clan   clan

      clan   clan   clan   clan

          clan   clan   clan
```

---

[77] Imm, C.
[78] "The Nomadic Tribes of Arabia"

So all of the clans were surrounded by enemies and were constantly either starting wars or being attacked. The clans that survived such a situation came up with a solution: polygamy!

> Muslims have justified multiple marriage for over a millennium."[79]

How could sanctioning polygamy help a clan survive? In two ways, one genetic and one social.

In a polygamous society, the married men (who are often better adapted in terms of wealth and/or power) have a greater chance to pass their genes on to the next generation than the unmarried males. This keeps honing the gene pool of the group. Secondly, it results in a lot of unmarried men who are ready to fight the unmarried men in the neighboring clans.

According to the Koran:

> Muslims are not allowed by any way to have physical relations with the opposite sex outside marriage.... If a man or a woman is forbidden from marriage for any a certain reason, he/she should observe abstinence.[80]

But how to convince men who will never have the opportunity to marry, go along with this deal? In two ways:

---

[79] Johnson, H.
[80] "Sex Outside of Marriage in Islam"

1) Convincing these young men that it is their duty to fight, and
2) Convincing them that if they die, they will be greeted in heaven by 72 virgins!

One of the canonical *hadiths* (statements attributed to Muhammad that mainstream Sunni Islam acknowledges as true) which all jihadi organizations regularly invoke depicts Muhammad saying: *The martyr [shahid, one who dies fighting for Islam] is special to Allah. He is forgiven from the first drop of blood [that he sheds]. He sees his throne in paradise. ... And he will copulate with seventy-two houris*[81][82]

And that's how you produce terrorists and convince them to commit terrorism.

Conclusion to Part Two
It's not nice to ignore Mother Nature!

Socialism is a "Potemkin Village" (a construction whose purpose is to provide an external façade to a situation, to make people believe that the situation is better than it actually is).

Democrats have convinced voters that the other side is evil and that people vote for them, everything will be fair and fine. Then four years later they make the same (false) promises.

---

[81] extremely beautiful young women
[82] Ibrahim, R.

Socialism is a drug. When you first get it, it feels great, but then you get addicted to it and need more and more of it.

Socialism is a "Ponzi scheme" ("a type of fraud in which the people who run the scheme tell customers they will invest their money. And for early investors, it is a good deal because the fraudsters pay the first customers with money paid by later customers. However, in the long run, the fraudsters pocket the money. And it is the same with socialism, except, instead getting more money from suckers, the socialists *borrow* money that will have to be paid by future generations.

The result of their Ponzi Scheme is that the U.S. national debt is now about *36 trillion dollars*! That works out to *$100,000* owed to the government by every person in the country! We are leaving a debt to our country that our descendants will never be able to pay off!

But, except for "climate change," people on the left don't care about the future; they are more interested in feeling sorry for and feeling good about themselves (and superior to anyone who disagrees with them). (Someone said that Bill Clinton made it all right for people to feel good about themselves without doing any good.)

As for socialist politicians (AKA, the Democratic Party), all they care about is getting reelected in any way possible so they can become richer and more powerful.

The rules of life have been evolving for billions of years, and socialism is not part of them.

Let freedom ring (and let nature take its course)!

**Part Three:
Conclusion**

Life — What a Concept! And What a Miracle!

"It is not the strongest of the species that survives, nor the most intelligent, but the one most responsive to change."

Charles Darwin

Nowadays people think they know the "theory of evolution," but they do not *believe in* it. They do everything they can to fight against the laws of evolution but, as they should know, "It's not nice to try to fool Mother Nature."

There is an old saying: "Though the mills of God grind slowly, they grind exceeding fine." For the discussion here, we can replace the word "God" with "Nature."

So, if you believe in God, his directive is: "Be fruitful and multiply." However, if you are an atheist, then you should abide by Nature's directive: Be fruitful and multiply." It's as simple as that.

As we go about our lives, which are exceedingly short in geologic time, there is a force that is shaping life not only over human lifetimes but over hundreds, thousands, millions, even billions of years, and that force is *evolution*.

And the important thing to remember is that individual organisms do not evolve or adapt; without outside interference*, groups* evolve/adapt when allowed to do so. The mechanism by which this occurs is that the individual organisms which are better adapted to the environment in which they find themselves have a better chance of surviving and passing their genes on to more offspring than the less well adapted organisms do.

Now you may be thinking that it would be unbearably cruel to "let nature take its course." But what you also need to remember is: 1) Your feelings of sympathy and philanthropy for your fellow human beings are, in a sense, an anachronism because they evolved in the past to make people help other people *with whom they shared the most genes*, and 2) Mother Nature *always wins* in the long run. Like a Ponzi scheme, circumventing the laws of Nature work in the short but must be "paid off" in the future by our descendants.

This fact is hard to accept. It seems so cruel, so hard to accept, so...unwoke! Unfortunately, the lives of organisms are not supposed to be easy because (as explained above) nobody is perfectly adapted to their environment, the environment their genes were adapted to may no longer exist, and, unfortunately, for a group of organisms to adapt and survive in the everchanging environment, some organisms will succeed with less effort than other organisms. (But also remember that certain genes that are not well adapted to the present environment may get lucky and be better adapted in the future when the environment "moves in their direction.")

A *Simple Grandeur*

Darwin's biological science of human nature challenged the utopian vision of the left by denying human perfectibility and suggesting that social reform would always be constrained by the limitations of human nature.[83]

---

[83] Dowd, M.

As Darwin himself wrote:

There is a simple grandeur in the view of life with its powers of growth, assimilation and reproduction, being originally breathed into matter under one or few forms, and that whilst this our planet has gone circling according to fixed laws, and land and water, in a cycle of change, have gone on replacing each other, that from so simple an origin, through the power of gradual selection of influential changes, endless forms most beautiful and most wonderful have been evolved.[84]

Or as Shakespeare famously wrote in *Hamlet*:

What a piece of work is a man! How noble in reason! How infinite in faculty! In form and moving how express and admirable! In action how like an angel! In apprehension how like a god! The beauty of the world! The paragon of animals!

Dear reader, *you* are one of those animals! Whether we have evolved according to the rules of Nature or God made us in his image, you are (as far as we know) the most advanced life form in the universe! *You* are the paragon of animals!

---

[84] "London Years"

And remember: you come from a *long* line of winners that goes back literally *billions* of years for, without exception *every single one of your ancestors survived long enough to have at least one offspring!*

I happen to believe that we evolved according to the rules of nature, but that means that every one of us contains in our genes the wisdom that has been gained through the struggle and pain of our ancestors over hundreds of millions—even billions—of years. This knowledge acquired through untold pain and hardship influences us in some ways obvious, in some ways not so obvious, and in some ways that are invisible to us.

Trust (or at least have faith in) that wisdom! In a way, your ancestors are looking out for you, saying, "Listen, this worked for me. It might work for you." (And if you're Jewish, they might add, "Listen—it couldn't hoit!")

Life may not have *great* meaning but it does have *some* meaning. It is up to you to find (or create) it. And even if you cannot create great meaning in your life, you can live a meaningful life just by passing the genetic inheritance you received from your ancestors down to your descendants, or by helping others do so, or at least helping them to find acceptance, gratitude, purpose, meaning, compassion and faith in their lives.

"Faintly trust the larger hope"

I have some bad news for you:

You can't win! You can't beat Mother Nature! Resistance is futile—you have *already* been absorbed into Nature's system! Your only choices are to fight

God/Mother Nature and lose or go along with him/her in your own most creative and dignified way.

Whether you believe in God or Nature, accept that you have been chosen by a Higher Power to carry out an important mission. Take responsibility for—and pride in—that mission and have faith that your life will have meaning if you try to fulfill it.

Or, as Tennyson wrote ("In Memoriam A. H. H."):

Are God and Nature then at strife,
That Nature lends such evil dreams?
So careful of the type she seems,
So careless of the single life;
That I, considering everywhere
Her secret meaning in her deeds,
And finding that of fifty seeds
She often brings but one to bear,
I falter where I firmly trod,
And falling with my weight of cares
Upon the great world's altar-stairs
That slope thro' darkness up to God,
I stretch lame hands of faith, and grope,
And gather dust and chaff, and call
To what I feel is Lord of all,
And faintly trust the larger hope.

In the final analysis, what can we mortals do but *faintly trust the larger hope?*

So what's it all about?
A New Way of Looking at "Life" (and Your Life!)

Sometimes the more you know about a situation, the more depressing it seems.

For the scientist Richard Dawkins,

[I]ndividual human beings simply function as "vehicles" for the replicators' survival. Arguing a similar vein to Dawkins, E. O. Wilson says of the genes that "the individual organism is only their vehicle, part of an elaborate device to preserve and spread [the genes] with the least possible perturbation." Simply stated, human creatures do not survive, but their genes can live on and on. Individual creatures come and go, but the genotype endures.[85]

If life is just about a bunch of molecules randomly building bodies to pass on copies of themselves, how can there be a meaning to life?

Good question, and a perplexing one, especially if you are looking for meaning in your life.

But, there is bright side to this conundrum:

If you were put here as the result of the struggles of billions of your ancestors (going all the way back to one-celled animals), then you should value your life because *it is indeed very "valuable"*!

---

[85] Miller, A.S. & Kanazawa, S.

Life is like being a runner in a torch relay

Think about the Olympic torch relay. Each participant helps the flame "survive" from its starting point to its endpoint where it will be used to light the Olympic torch at place where there Games will be, but….

Do the chemicals that keep the flame alive survive the relay? No! Do the runners finish the race? No. So what survives? *The energy that keeps the flame alive* when it is passed to a new torch.

Life is like that. Each of us helps pass the "life energy" (the genetic code) that our ancestors kept alive over millennia to our descendants, who will hopefully do the same.

In addition, if you realize that your primary job in life is to stay alive and help your genes (and the genes that other people who are genetically related to you) continue to exist to be passed into new beings, doesn't that give you a sense of importance to be chosen to attempt that noble task?

It can also simplify things and can help focus your mind on — and direct more energy to — those things that are really important in your life (and to "life," in general).

Is a loss by your favorite sports team really a thing to get sad or angry about? Does what strangers think about you really matter? Do problems that you will never face and can do nothing about really matter? Does it really matter in the larger scale of things if you show up at a costume party and someone else is wearing the exact same expensive costume that you are wearing?

If you can focus only on the reason you were created and the things that you can control (or that can control you) and forget about all that other nonsense, you can lead a noble, satisfying simpler life.

As Ghandi said:

Whatever you do in life will be insignificant but it is very important that you do it because you can't know. You can't ever really know the meaning of your life. And you don't need to. Every life has a meaning, whether it lasts one hundred years or one hundred seconds. Every life, and every death, changes the world in its own way. You can't know. So don't take it for granted. But don't take it too seriously. Don't postpone what you want. Don't leave anything misunderstood. Make sure the people you care about know. Make sure they know how you really feel. Because just like that... It could end."[86]

---

[86] "Ghandi: Quotes"

Final Note from the Author

If you have made it all the way to here, congratulations! You deserve ten out of ten on the "willing to consider new ideas" scale.

I hope you enjoyed this book. Even if it didn't make you change your ideas, I hope it made you got you to look at "Life" (and your life) from a new perspective.

Please excuse any mistakes, omissions, repetitions or contradictions in the above. I' only human!

Keep believing in yourself and nature and remember you are special (in fact, *unique* — there has never been anyone like you in the Universe and there never will be in the future), that you were created for an achievable purpose, and that what you have learned in this book is not all bad. You have been given the abilities to laugh and love and be creative and feel happy for a reason!

Richard Showstack

s

136

Sources

Aird, W. "William Harvey on the Blood." (Accessed Sept. 4, 2024, from https://www.thebloodproject.com/william-harvey-on-the-blood/#:~:text=doctrine%20once%20sown%20strikes%20deep,the%20candour%20of%20cultivated%20minds.

"The Allegory of the Cave." *Plato: Education and the Value of Justice.* (Accessed July 28, 2009, from http://www.philosophypages.com/hy/2h.htm)

"Are Humans Still Evolving? 'Maybe More Rapidly Than Ever,' Says Scientist." Newsweek. Accessed Sept. 18, 2024, from: https://www.newsweek.com/humans-evolving-rapidly-ever-scientist-evolution-genetics-1852884

Arnhart, L. (2005). *Darwinian Conservatism.* Charlottesville, VA.: Imprint Academic.

"Biology: Study of Life." (Dec 7, 2023) (Accessed Oct. 1, 2014 from: https://quizlet.com/study-guides/biology-study-of-life-e953b7b9-f12a-4e44-846d-f68765472aca

"Birth rate in the United States in 2019, by household income." Statista. (Accessed Sept. 13, 2024, 2019, from https://www.statista.com/statistics/241530/birth-rate-by-family-income-in-the-us/)

"Births: Final Data." National Vital Statistics Reports. (Accessed Sept. 17, 2024 from: https://www.cdc.gov/nchs/data/nvsr/nvsr73/nvsr73-02.pdf

Boyd, C. A. (2007). *A Shared Morality: A Narrative Defense of Natural Law Ethics.* Grand Rapids, MI: Brazos Press.

Bryozoa. *Wikipedia.* (Accessed Aug. 6, 2019, from https://en.wikipedia.org/wiki/Bryozoa)

Callaway, E. (September 10, 2008). "Superstitions evolved to help us survive." *NewScientist.* (Accessed July 20, 2009,

from http://www.newscientist.com/article/dn14694-
superstitions-evolved-to-help-us-survive.html)

Chand, S. (Sept. 2, 2009). "We are all mutants say scientists."
BBC. (Accessed September 2, 2009, from
http://news.bbc.co.uk/2/hi/science/nature/8227442.stm)

Clough, A. "China Faces Consequences of the One-Child Policy.
(Feb. 14, 2024) Accessed Sept. 16, 2024 from:
https://providencemag.com/2024/02/china-faces-the-
consequences-of-one-child-
policy/#:~:text=It's%20become%20clear%20that%20the,old
%20brackets%2C%20outnumbering%20younger%20ones.

"Continents Of The World By Total Fertility Rates." Accessed
Sept. 15, 2024 from:
https://www.worldatlas.com/articles/continents-of-the-
world-by-total-fertility-rates.html

"Countries with the highest fertility rates 2024." (August, 2024)
Accessed October 2, 2024, from:
https://www.statista.com/statistics/262884/countries-with-
the-highest-fertility-rates/

Cox, J. (April 12, 2019). "Alan Greenspan says economy will
start to fade 'very dramatically' because of entitlement
burden." *CNBC*. (Accessed Aug. 8, 2019, from
https://www.cnbc.com/2019/04/12/alan-greenspan-says-
economy-will-start-to-fade-out-because-of-growing-us-
entitlement-burden.html)

Darwin, C. (1859). *On the Origin of Species*. (Accessed July 20,
2009, from
http://www.csuchico.edu/~curbanowicz/DarwinDayCollectio
nOneChapter.html)

"Did Evolution Make Us Cancer Prone?" (July 3, 2009).
ScienceDaily. (Accessed July 20, 2009, from
http://www.sciencedaily.com/releases/2009/07/09070211421
0.htm)

Dowd, M. (2007). *Thank God for Evolution! How the Marriage of Science and Religion Will Transform Your Life and Our World.* San Francisco: Council Oak Books.

"Europeans too selfish to have children, says Chief Rabbi." (January 13, 2010). *The Telegraph.* (Accessed January 13, 2008, from http://www.telegraph.co.uk/news/newstopics/religion/65077 82/Europeans-too-selfish-to-have-children-says-Chief-Rabbi.html)

"Evolution of Primates." *Wikipedia.* (Accessed Aug. 6, 2019 from https://en.wikipedia.org/wiki/Evolution_of_primates)

"Fertility rate in each continent and worldwide, from 1950 to 2024." Accessed Oct. 2, 2024, from: https://www.statista.com/statistics/1034075/fertility-rate-world-continents-1950-2020/

Forrest, W. *The Conversation.* (Nov. 7, 2011) "Marriage helps reduce crime." (Accessed Oct. 8, 2024, from: https://theconversation.com/marriage-helps-reduce-crime-3576#:~:text=Other%20researchers%20have%20observed%20similar,have%20had%20long%20criminal%20careers)

"Ghandi: Quotes." Goodreads. (Accessed Sept. 18, 2024, from: https://www.goodreads.com/quotes/7960679-whatever-you-do-in-life-will-be-insignificant-but-it#:~:text=You%20can't%20ever%20really,world%20in%20its%20own%20way.

Giberson, K. W. (2008). *Saving Darwin: How to Be a Christian and Believe in Evolution.* New York: HarperCollins.

Hannan, D. (April 18, 2009). "If you pay people to be poor, you'll never run out of poor people." *Telegraph.co.uk.* (Accessed February 22, 2010 from http://blogs.telegraph.co.uk/news/danielhannan/9561761/If_you_pay_people_to_be_poor_youl l_never_run_out_of_poor_people/)

Hayden, T. "What Darwin Didn't Know: Today's scientists marvel that the 19th-century naturalist's grand vision of evolution is still the key to life." (February, 2009). *Smithsonian.com.* (Accessed Aug. 6, 2019, from https://www.smithsonianmag.com/science-nature/what-darwin-didnt-know-45637001/)

Ibrahim, R. (December 12, 2018). "Killing and Dying for the 'Houris': Islam's Heavenly Whores." *Middle East Forum.* (Accessed Oct. 8, 2024, from: https://www.meforum.org/killing-and-dying-for-the-houris-islam-heavenly

Imm, C. (March 3, 2023) "Untilled Earth: The Causes and Dangers of America's Single Young Men." *Midwestern Citizen.* Accessed Oct. 8, 2023, from: https://midwesterncitizen.substack.com/p/untilled-earth-the-causes-and-dangers

"Income and fertility." Wikipedia. (Accessed Sept. 15, 2024 from: https://en.wikipedia.org/wiki/Income_and_fertility#:~:text=There%20is%20generally%20an%20inverse,born%20in%20any%20developed%20country.

"Is nearsighted genetic?" *All About Vision.* (Accessed Sept. 19, 2024, from: https://www.allaboutvision.com/conditions/myopia-faq/is-being-nearsighted-genetic.htm#:~:text=And%20genetics%20definitely%20play%20a,to%20myopia%20and%20refractive%20error.

Johnson, H. "There are Worse Things Than Being Alone: Polygamy in Islam, Past, Present, and Future." (April, 2005). *William & Mary Journal of Women and the Law.* Accessed Oct. 8, 2024, from: https://scholarship.law.wm.edu/cgi/viewcontent.cgi?article=1134&context=wmjowl

Joyce, R. (2006). *The Evolution of Morality*. Cambridge, MA: MIT Press.

Kaplan, K. (June 25, 2009). "Study finds humans still evolving, and quickly." *Los Angeles Times*. (Accessed July 20, 2009, from http://www.latimes.com/news/science/la-sci-evolution11dec11,1,3366709.story)

Kirchner, L. (July 15, 2013). "Marriage May Calm a Criminal Impulse in Men." Pacific Standard. Accessed Oct. 8, 2024, from: https://psmag.com/social-justice/marriage-may-calm-a-criminal-impulse-in-men-62504/

Krugman, P. (January 11, 2010). "Learning From Europe." *The New York Times*.
(Accessed January 13, 2010 from http://www.nytimes.com/2010/01/11/opinion/11krugman.html?em=&pagewanted=print)

"Life." *Wikipedia.* (Accessed August 5, 2019 from https://en.wikipedia.org/wiki/Life)

"List of countries by total fertility rate." (Accessed Sept. 15, 2024, from:
https://en.wikipedia.org/wiki/List_of_countries_by_total_fertility_rate

"London Years: Darwin's scientific manuscripts in the aftermath of the Beagle voyage." American Museum of Natural History. (Accessed Sept. 18, 2024, from: https://www.amnh.org/research/darwin-manuscripts/edited-manuscripts/evolution-papers/creating-the-origin/london-years

Masci, D. (1). (Feb. 6, 2019). "Darwin in America: The evolution debate in the United States" *Pew Research Center: Religion & Public Life.* (Accessed Aug. 6, 2019 from https://www.pewforum.org/essay/darwin-in-america/)

Matthews, M. (Feb. 12, 2019). "Obamacare Was Supposed to Lower the Death Rate—It Didn't." *IPI (Institute for Policy Innovation).* (Accessed Aug. 8, 2019, from https://www.ipi.org/ipi_issues/detail/obamacare-was-supposed-to-lower-the-death-rate-it-didnt)

Miller, A.S. & Kanazawa, S. (2007). *Why Beautiful People Have More Daughters.* New York: Penguin Books.

(2008). *Only a Theory: Evolution and the Battle for America's Soul.* New York: Viking Penguin.

"Mother's Educational Level Influences Birth Rate," a summary of the article, "Birth and Fertility Rates by Educational Attainment: United States, 1994," by T. J. Mathews and Stephanie J. Ventura. (Accessed December 18, 2009, from: http://library.adoption.com/articles/ mothers-educational-level-influences-birth-rate.html)

"Nearsightedness Rates Are Soaring. Here's Why." Scientific American. (Accessed Sept. 19, 2024, from: https://www.scientificamerican.com/article/nearsightedness-rates-are-soaring-heres-why/#:~:text=Optometry%20researchers%20estimate%20that%20about,health%20care%20costs%20are%20huge.

"The Nomadic Tribes of Arabia." *World Civilization.* (Accessed Oct. 8, 2024, from: https://courses.lumenlearning.com/suny-hccc-worldcivilization/chapter/the-nomadic-tribes-of-arabia/#:~:text=Society%20was%20patriarchal%2C%20with%20inheritance,viewed%20as%20outsiders%20or%20enemies.

On the Motion of the Heart and Blood in Animals: William Harvey. (Accessed Sept. 210, 2014, from: http://web.ecs.baylor.edu/faculty/newberry/myweb/gtx%20os%20files/f05%20student%20pages/morehouse-basinger/Harvey%201.htm#:~:text=I%20tremble%20lest%20I%20have,the%20candour%20of%20cultivated%20minds."

Pinker, S. (2002). *The Blank Slate: The Modern Denial of Human Nature*. New York: Viking Penguin.

Rudy, M. (September 23, 2024) "Unsustainable." (Retrieved Sept. 24, 2024, from: https://www.foxnews.com/health/growing-uk-health-care-crisis-millions-patients-waiting-care-data-shows

Sampson, J., et al. (November 3, 2006). "Does Marriage Reduce Crime? A Counterfactual Approach To Within-Individual Causal Effects." (Accessed Oct. 8, 2024, 2024, from: https://scholar.harvard.edu/files/sampson/files/2006_criminology_laubwimer_1.pdf

Schoemaker, J. (1991) "Social Class as a Determinant of Fertility Behavior: The Case of Bolivia." In: *Proceedings of the Demographic and Health Surveys World Conference*. pp. 73-88. Institute for Resource Development/Macro International, Demographic and Health Surveys [DHS]: Columbia, Maryland.

"The Selfish Gene." Wikipedia. (Accessed Sept. 18, 2024, from: https://en.wikipedia.org/wiki/The_Selfish_Gene)

"Sex Outside of Marriage in Islam." *About Islam*. Accessed Oct. 8, 2024, from: https://aboutislam.net/counseling/ask-the-scholar/crimes-penalties/how-does-islam-view-sex-outside-marriage/

Smith, C. M. & Sullivan, C. (2007). *The Top 10 Myths About Evolution*. New York: Prometheus Books.

Sowell, T. (2007). *A Conflict of Visions: Ideological Origins of Political Struggles (Revised Edition)*. New York: Basic Books.

"Speciation." *Scitable by Nature Education*. (Accessed Aug. 6, 2019 from https://www.nature.com/scitable/definition/speciation-183)

"Species & Speciation." *Khan Academy.* (Accessed Aug. 6, 2019, from https://www.khanacademy.org/science/biology/her/tree-of-life/a/species-speciation)

Than, K. "All Species Evolved From Single Cell, Study Finds." *National Geographic.* (Accessed Aug. 6, 2019 from https://news.nationalgeographic.com/news/2010/05/100513-science-evolution-darwin-single-ancestor/

"The Evolution of Evil." (Accessed Sept. 24, 2024, from: https://sites.google.com/view/psychology-of-evil/07-the-evolution-of-evil

"Third World population growth at record high." (Accessed Oct. 2, 2024, from: https://pubmed.ncbi.nlm.nih.gov/12288263/#:~:text=Women%20in%20the%20richer%20countries,to%20stabilize%20world%20population%20size.

"The 20 countries with the lowest fertility rates in 2024." Accessed Oct. 2, 2024, from: https://www.statista.com/statistics/268083/countries-with-the-lowest-fertility-rates/#statisticContainer

"Tomasello, T." "The Origins of Human Morality: How we learned to put our fate in one another's hands." 2018. Scientific American. Accessed Sept. 18, 2024, from: https://www.scientificamerican.com/article/the-origins-of-human-morality/

"Total Fertility Rate." *Wikipedia.* (Accessed Aug. 8, 2019, from https://en.wikipedia.org/wiki/Total_fertility_rate)

Yamaguchi, Y. "Japan birth rate hits record low amid concerns over shrinking and aging population." (June 2, 2023) Accessed September 16, 2024, from: https://apnews.com/article/japan-birth-rate-record-low-population-aging-ade0c8a5bb52442f4365db1597530ee4

Wilson, D. S. (2007). *Evolution For Everyone: How Darwin's Theory Can Change the Way We Think About Our Lives.* New York: Bantam Dell.

"The World of Darwin: The Passage of Time." *Exploring Life @ Bio.dot.Edu.* (Accessed Aug. 7, 2019, from http://www.brooklyn.cuny.edu/ bc/ahp/LAD/C20/C20_Circum.html)

Wright, R. (1994). *The Moral Animal—Why We Are the Way We Are: The New Science of Evolutionary Psychology.* New York: Vintage Books.

―――――――――――――――

www.ingramcontent.com/pod-product-compliance
Lightning Source LLC
Chambersburg PA
CBHW071505220526
45472CB00003B/926